AF365179

Workshop Practice Manual
Sixth Edition

Workshop Practice Manual
Sixth Edition

Prof. K. Venkata Reddy

Mechanical Engineering Deptartment
Yogananda Institute of Technology & Science
Tirupati - 517 520

BS Publications
A unit of **BSP Books Pvt., Ltd.**

4-4-309/316, Giriraj Lane, Sultan Bazar,
Hyderabad - 500 095
Phone : 040 - 23445605, 23445688

Published by :

 **BS Publications**

A unit of **BSP Books Pvt., Ltd.**

4-4-309/316, Giriraj Lane, Sultan Bazar,
Hyderabad - 500 095
Phone : 040 - 23445605, 23445688
e-mail : info@bspbooks.net

ISBN : 978-93-5230-041-9 (HB)

Preface to Sixth Edition

The curriculum of professional degree courses stresses more on practical skills based on the theory taught in most of the subjects.Now-a-days the importance of practicals is reduced drastically by making the university examinations in practicals a mere formality rather than a test of practical skill and knowledge of the students. This requires review if the engineers are to be accepted readily by industrialists.

This book on workshop practice as a manual, meets the requirement in that direction. A chapter on IT workshop is added in the revised edition to induce interest on hardware and assembly/disassemblyof computers, which is required for all branches of engineering.

I thank M/S B S Publications for bringing out this manual in its present form at a short notice.

-Author

Preface to First Edition

Practice makes a man perfect. All engineering students studying various courses in different branches of engineering are to acquire both theoretical and practical knowledge so as to serve the society in improving the living standards.

The workshop practicals in first year, common to students of all brances of engineering at diploma and degree level is intended to impart basic knowledge of various hand tools and their use in different sections of manufacturing. They include Carpentry, Fitting, Electrical wiring, Blacksmithy, Sheet metal, Welding, Foundry, Machine shop and Plumbing. The intention of All India Council for Technical Education (AICTE), insisting on the provision of these workshop facilities apart from other advanced laboratories for individual branches is therefore very clear.

The author with his vast experience as workshop superintendent has identified the difficulties faced by the students in understanding the importance of the practicals and writing the records in the way it should be. To overcome the difficulties and also to motivate the students for better standards of practical approach, this material is designed as source of information. Worksheets are added for use in the practical class and replaces the observation book.

An attempt is made to motivate the students on human, moral and spiritual values which is the need of the day by presenting a few adages on each page. This, the author believes will help the students to inculcate positive mental attitude and develop their personality and leadership qualities. More than the IQ of a person the need of the day is E.Q. (Emotional Quotient) and T.Q (Transmission Quotient or Communication Skills).

-Author

Contents

Contents (xi)

Making the technology survive
(Source : The Engineer in you)

Engineering is the conversion of the forces of nature for the fulfilment of man's needs. Therefore, it is also the Engineers responsibility to see that development does not result in un-safe (polluted) world. The biggest enemy today over and above the environmental pollution is the mental pollution of the people at the helm of administration in all fields. Save the world from unsafe hands.

DEPARTMENT OF MECHANICAL ENGINEERING

CERTIFICATE

Certified that this is the bonafide record of work done in First year

workshop practicals, by Name : ....

Class : .... Roll No.: .... during the year 201 - 201

Signature of Staff   **Signature of HOD**

Internal Examiner   **External Examiner**

Index

Ex. No.	Date	Name of the Excercise	Page no.	Initials
1				
2				
3				
4				
5				
6				
7				
8				
9				
10				
11				
12				
13				
14				
15				
16				

Index

Ex. No.	Date	Name of the Excercise	Page no.	Initials
17				
18				
19				
20				
21				
22				
23				
24				
25				
26				
27				
28				
29				
30				
31				
32				

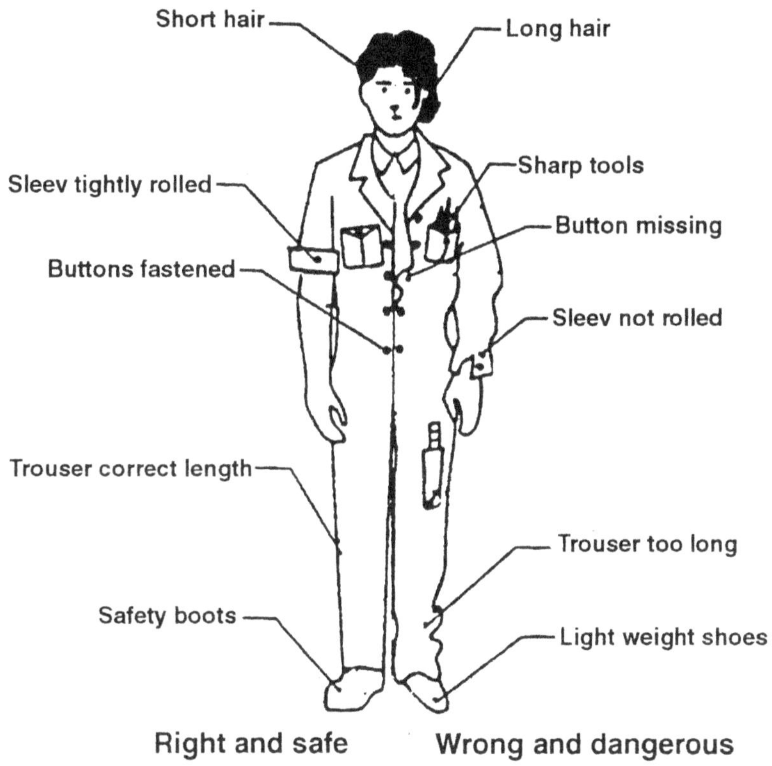

WORKSHOP DRESS

SAFETY RULES AND UNSAFE PRACTICES

GENERAL SAFETY RULES

Remember that "accidents do not occur, they are caused". With this in mind, strictly follow the general safety rules given below and safe practices indicated in brief under each section.

1. Safety first, work next.

2. Know your job and follow instructions.

3. Avoid wearing clothing that might catch, moving or rotating parts. Long sleeves of shirts, long hair,neck tie and jewellery are definite hazards in the shop.

4. Wear safety shoes. Do not wear canvas shoes; they give no resistance to hard objects dropped on the feet.

5. Keep the area around the machine or work clean.

6. Keep away from revolving work.

7. Be sure that all gaurds are in place.

8. One person only should operate the machine controls.

9. Use tools correctly and do not use them if they are not in   proper working condition.

10. Wear safety goggles when working in areas, where sparks or   chips of metal are flying.

11. Never (a) operate a machine unless you are authorised to do so. (b) Start a machine unless you know how to stop it. (c) Walk away and leave a machine running. d) Distract or interfere with any one, operating a machine.

12. Get to know who is in-charge of first-aid and where boxes are   placed and where the first aid can be found in case of emergency.

CHAPTER - 1

FITTING SHOP

INTRODUCTION

Manufacturing processes are broadly classified into four categories; (i) Casting processes, (ii) Forming processes, (iii) Fabrication processes, and (iv) Material removal processes.

In all these processes, components are produced with the help of either machines or manual effort. The attention of a fitter is required at various stages of manufacture starting from marking to assembling and testing the finished goods.

Working on components with hand tools and instruments, mostly on work benches is generally referred to as 'Fitting work'. The hand operations in fitting shop include marking, filing, sawing, scraping, drilling, tapping, grinding, etc.,using hand tools or power operated portable tools. Measuring and inspection of components and maintenance of equipment is also considered as important work of fitting shop technicians.

WORK HOLDING TOOLS

Bench Vice

The bench vice is a device commonly used for holding the work pieces. When the vice handle is turned in a clockwise direction the moving jaw forces the work against the fixed jaw. The greater the pressure applied to the handle, the tighter is the work held. The body of the vice is made of cast-iron. Hardened steel plates with serrations to ensure better gripping of the work are fixed on the faces of the two

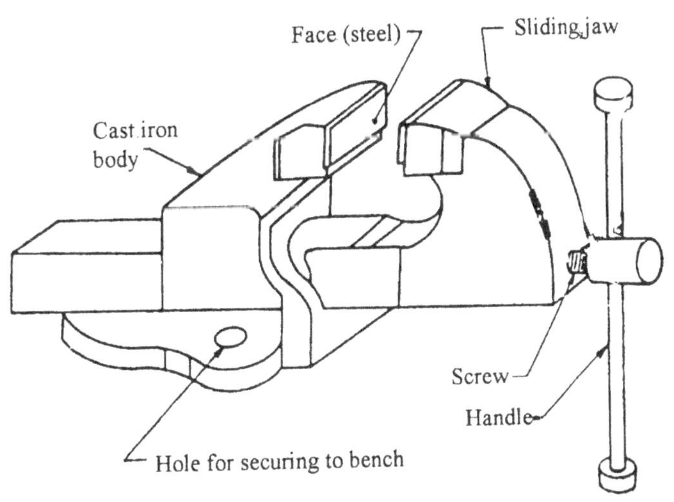

Bench Vice

jaws. Jaw caps made of soft material such as aluminium or galvanised iron (G.I) sheet are used to protect finished surfaces of the work gripped in the vice. Vices are specified by the maximum width that can be held or the maximum opening between the jaws, varying from 75mm to 300mm.

V-block with clamp

The V-block is a rectangular or square block with a V-groove on one or both sides, opposite to each other. The angle of the V is usually 90°. V-block with a clamp is used to hold cylindrical work securely, during marking of measurements or for measuring operations. Material: C.I or hardened steel. Size: 50 to 150mm.

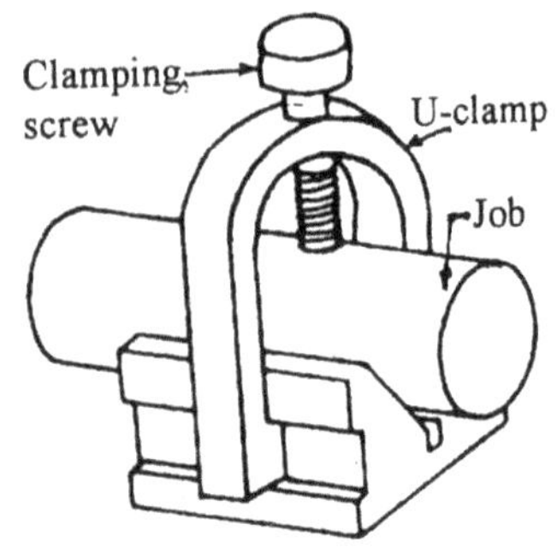

V - Block

Parallel Clamp

It is a simple screw clamp with parallel jaws to hold small jobs for working on them.

C - Clamp

This is used to hold work against an angle plate or V-block or any other surface, when gripping is required. It is also known as G-clamp.

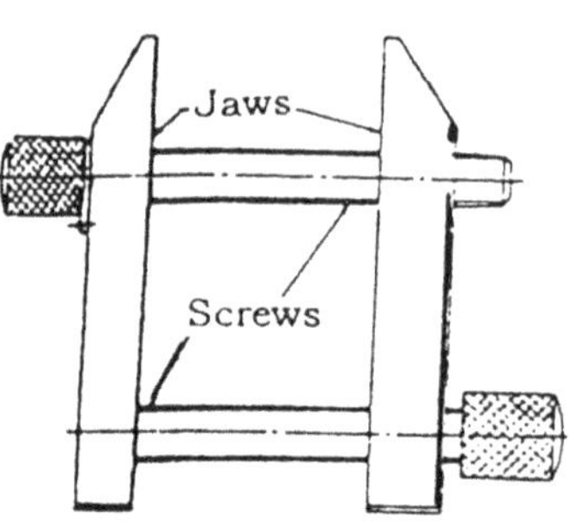

Parallel clamp

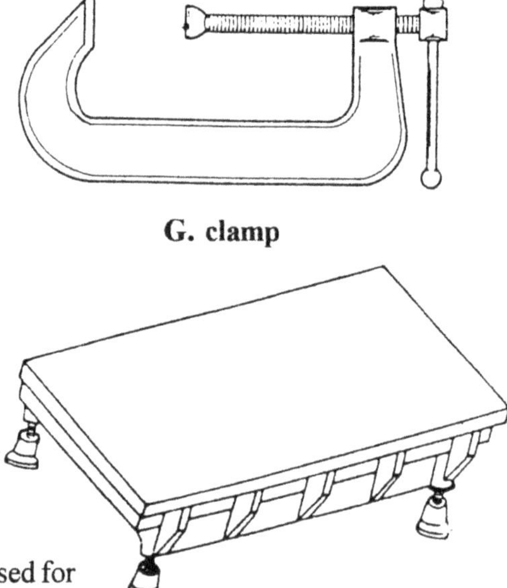

G. clamp

Marking And Measuring Tools

Marking table

A marking table is a heavily build cast iron table used for layout work on all sizes of jobs. This table provides a flat surface to mark lines with the help of height gauge, angle plate, V-block or surface gauge as per job requirements.

Marking table

Surface plate

The surface plate is used for testing the flatness of the work piece and other inspection purposes. It is also used for marking on small works. It is more precise in flatness than the marking table.

Surface plates are made of C.I. or hardened steel, ground and scraped to the required precision. Now-a-days surface plates made of special granite stone are manufactured in wide range of precision grades, colours and sizes. It is specified by length X width X height X grade. Example: 600 x 400 x 100 x grade A has a flatness upto 0.005mm.

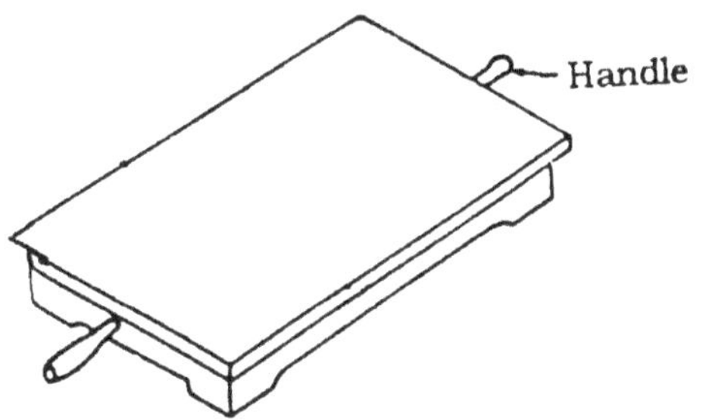

Surface plate

Angle plate

The angle plate is made of cast iron. It has two surfaces machined at right angles to each other. Plates and components which are to be marked out may be held against the upright face of angle plate to facilitate the marking or inspection.

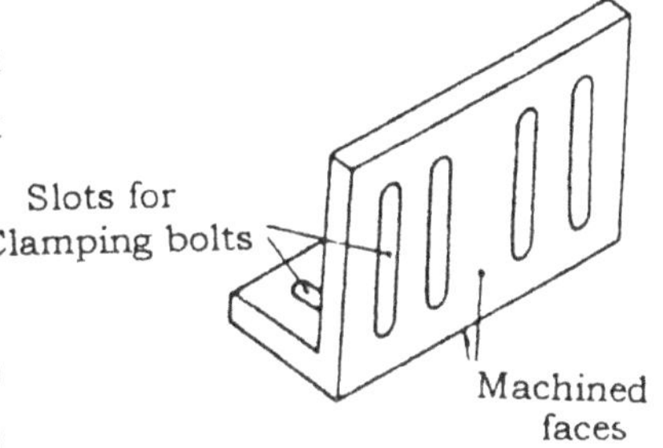

Angle Plate

Universal Scribing Block

This is used for scribing lines for layout work and checking parallel surfaces.

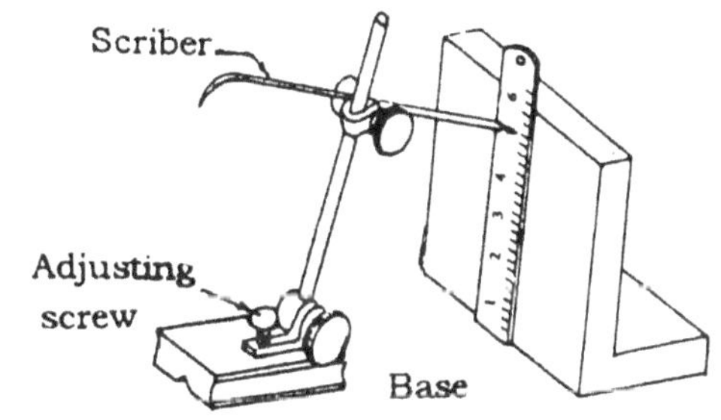

Universal scribing block

Try - Square

Try-square is used for checking the squareness of small works, when extreme accuracy is not required. The size of the try-square is specified by the length of the blade. Ex: 10cm, 30cm etc.

Scriber

A Scriber is a slender steel rod, used to scribe or mark lines on metal work pieces.

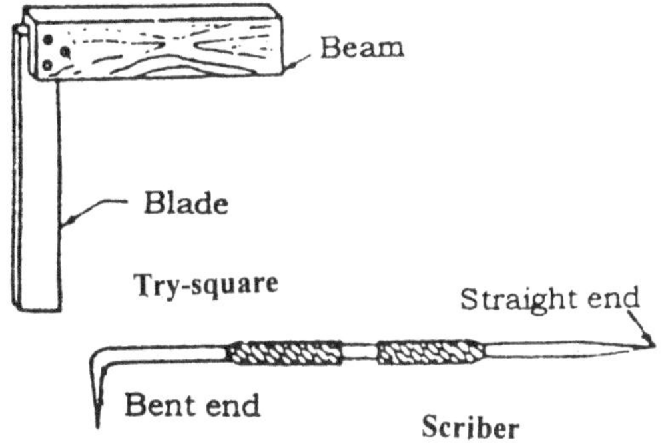

Combination set

A combinaton set consists of a rule, square head , centre head and a protractor. This may be used as a rule, a square, a depth gauge, for marking mitres (45 degrees), for measuring and marking angles. The rule in made of tempered steel with grooves.

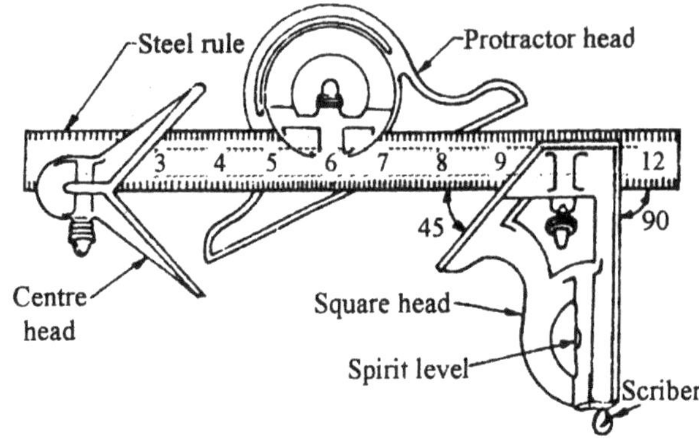

Combination set

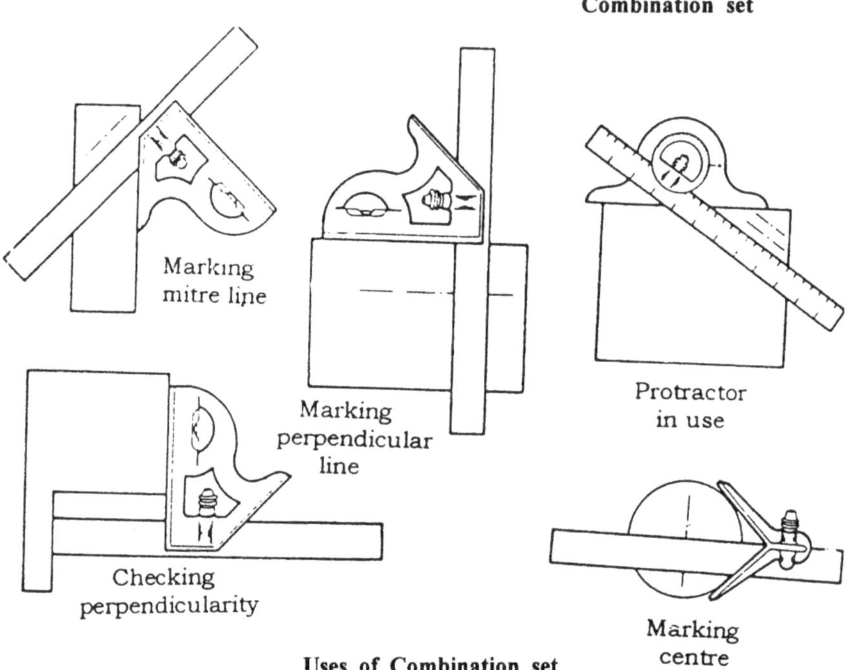

Uses of Combination set

Odd-leg caliper

This is also called 'jenny caliper' or 'hermaphrodite'. This is used for marking parallel lines from a finished edge and also for locating the centre of round bars. They are specified by the height of the leg upto the hinge point. Example: 100mm, 150mm etc.

Divider

This is used for marking circles, arcs, laying out perpendicular lines, bisecting lines, etc. Size ranges from 100 mm to 300 mm.

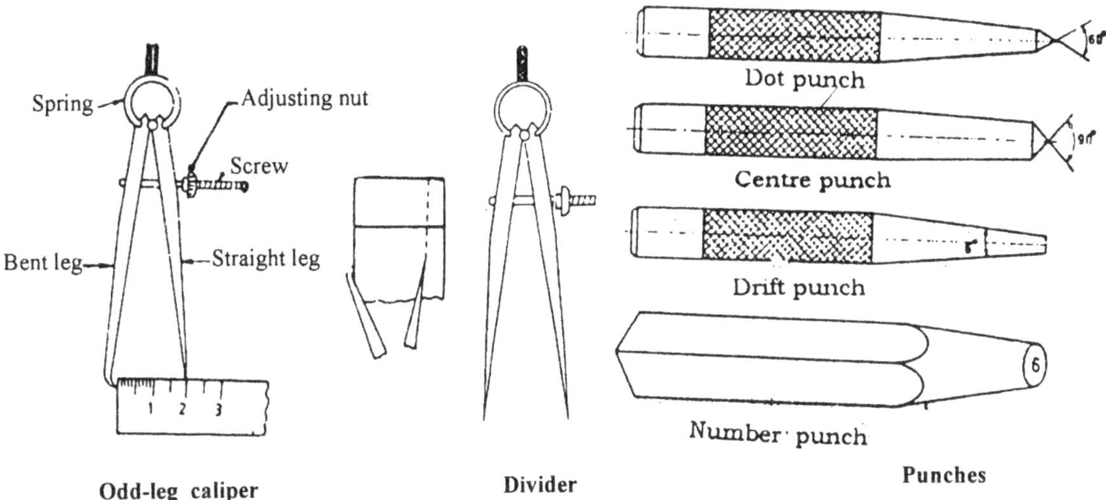

Dot punch

This is used to locate centre of holes and to provide a small centre mark for divider point etc. For this purpose, the punch is ground to a conical point having 60° included angle.

Centre punch

This is similar to the dot punch, except that it is ground to a conical point having 90° included angle. It is used to mark the location of the centre where holes are to be drilled. The centre punch mark facilitates easy location of the drill tip and centre accurately.

Drift punch

A drift punch is a long tapered tool used to align holes in two or more pieces of material that are to be joined together, so that bolts or rivets can be easily placed in the holes.

Letter punch

It has square body with a tapered end. At this end, a projection, corresponding to the replica of the letter to be marked is made. The letters used are A to Z, & totaling 27 numbers.

Number punch

It is similar to letter punch in construction but has numbers at its end. The numbers used are from 0 to 8 (six used as nine also). Punches are made of tool steel, hardened and tempered.

MEASURING TOOLS

Calipers

These are used with the help of steel rule to check outside and inside measurements. They are specified by the maximum length measured. Sizes vary from 100 mm to 300 mm.

Vernier calipers

These are used for measuring outside as well as inside dimensions accurately. It may also be used as a depth gauge. In the figure shown, 19 main scale divisions are divided into 20 equal parts in the vernier scale. Hence, least count of the vernier = 1 main scale division - 1 vernier scale division = 1-19/20 = 0.05 mm.

The size is specified by the maximum measurement it can make ranging from 150 to 300mm. The accuracy of the instrument depends on the least count, varying from 0.1 to 0.02. Other types of verniers include dial vernier, digital vernier with more accuracy etc.

Vernier height gauge

The vernier height gauge, clamped with a scriber, is shown in figure. It is used for layout work. An offset scriber is used when it is required to take measurements from the surface, on which the gauge is standing. The accuracy and working principle of the gauge are the same as those of the vernier caliper. The capacity of

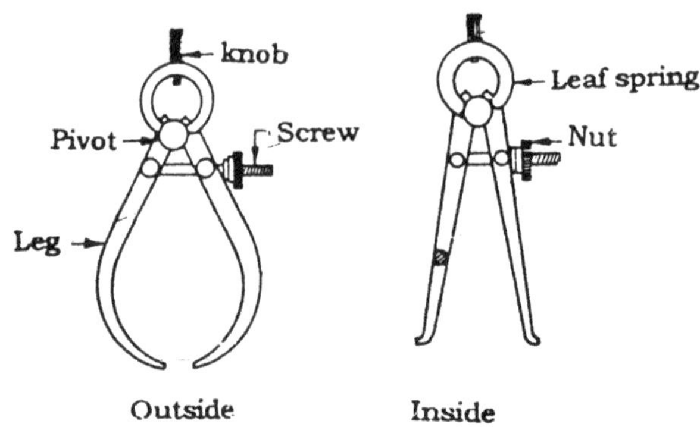

Spring calipers

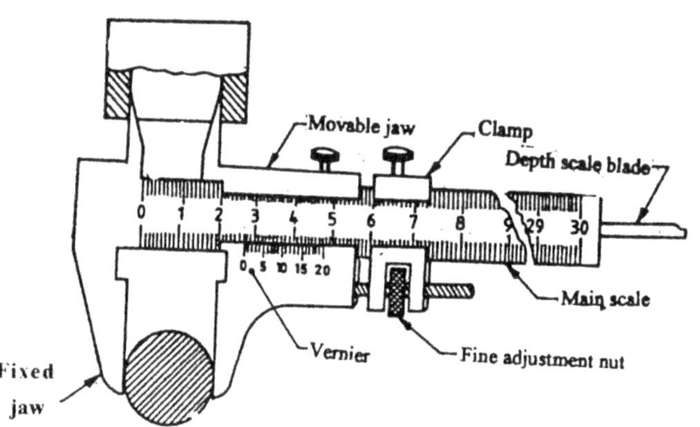

Vernier calipers

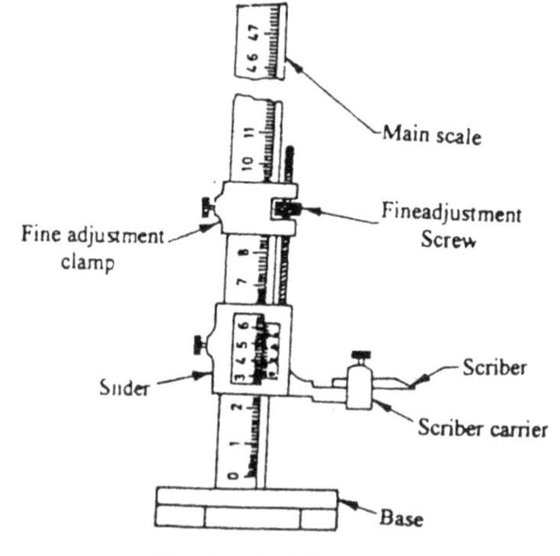

Vernier height gauge

the height gauge is specified by the maximum height it can measure. It varies from 150 mm to 1000 mm.

Outside micrometer

This is used for measuring external dimensions accurately. Figure shows a micrometer of 0 to 25mm range with an accuracy of 0.01mm. These are available in different ranges with interchangeable anvils varying from 0-25 mm to 2000 mm in sizes and 0.01 to 0.001 in accuracy. There are many types of micrometers designed for special purpose use. They include thread micrometers to measure thread dimensions, tube micrometers to measure wall thickness of tubes, etc.

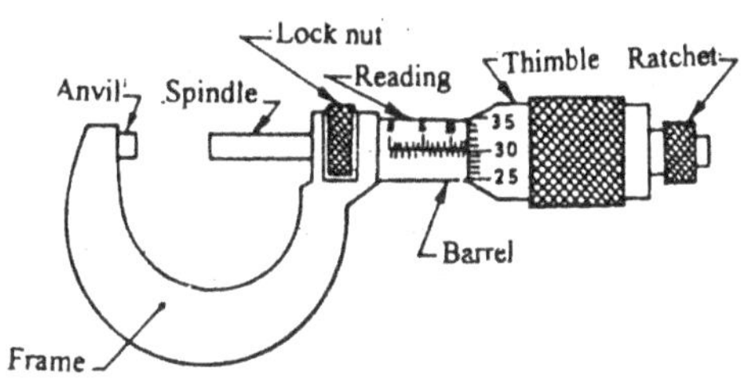

Outside micrometer

Inside micrometer

This is used to measure inside dimensions accurately. Figure shows an inside micrometer of range 25 to 150mm with extension rods. These are available in different ranges and accuracies.

Depth micrometer

It is designed to measure the depth of holes, slots, recesses etc. The working principle of this is similar to the outside micrometer. Its base is hardened ground and lapped to reduce wear. These are available upto a range of 300 mm and accuracy of 0.01 mm. In this the reading is taken from the thimble end (right to left), unlike the outside micrometer where reading is taken from left to right.

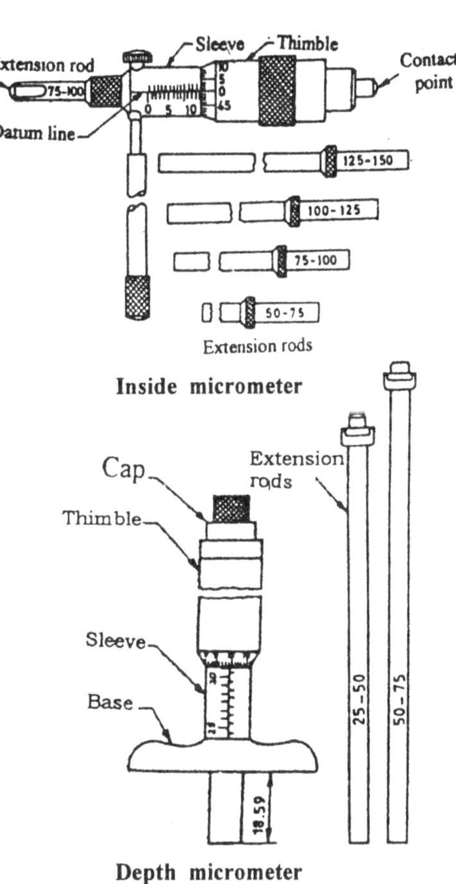

Depth micrometer

Feeler gauges

The thickness gauges or feeler gauges are a set of gauges consisting of thin strips of metal of varying thickness. They are widely used for measuring and checking bearing-clearance, adjusting tappets, spark plug gaps, and so on. The thickness varies from 0.05 to 0.5 mm.

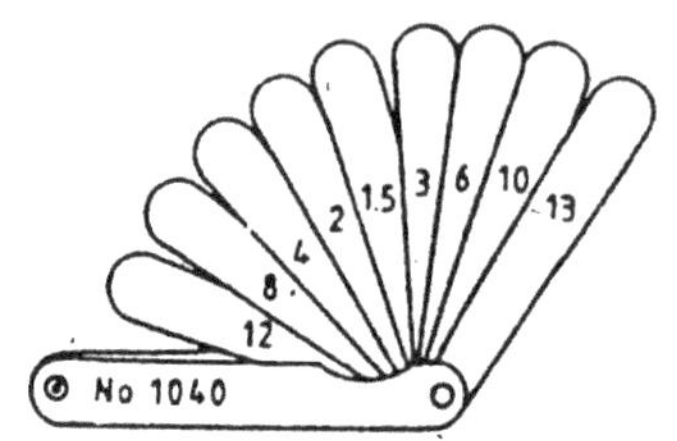

Feeler guage

Radius gauges

Also known as fillet gauges, these are of thin flat steel tool used for inspecting and checking, or laying out work having a given radius. Such a gauge is made of in sets of individual gauges for measuring concave (internal) or convex (external) radius.

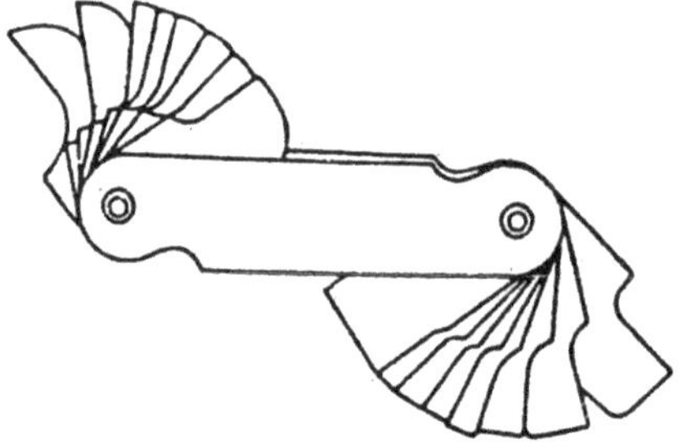

Radius guage

Screw pitch gauges

A screw pitch gauge is used for quickly determining the pitch of a threaded part or tapped hole. The gauge consists of a set of templates of teeth, each confirming to a standard pitch.

Drill gauge

Thin sheets with holes drilled accurately to the size marked are used as drill guages for easy selection and checking of drill size. This is very much useful when the drill size marked on the drill wears out over repeated usage. These guages are also available as stands for letter drills and number drills which are very small in size.

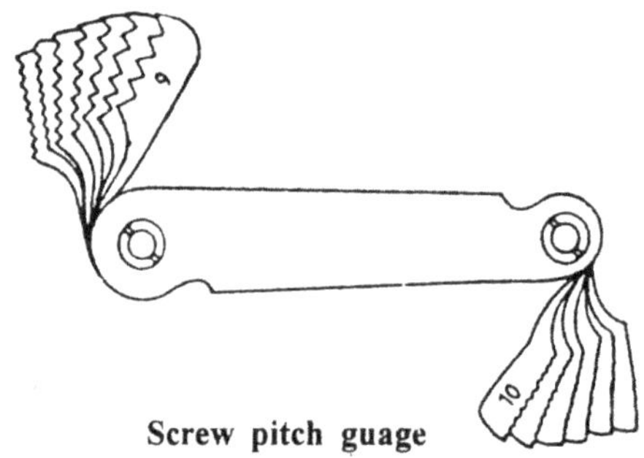

Screw pitch guage

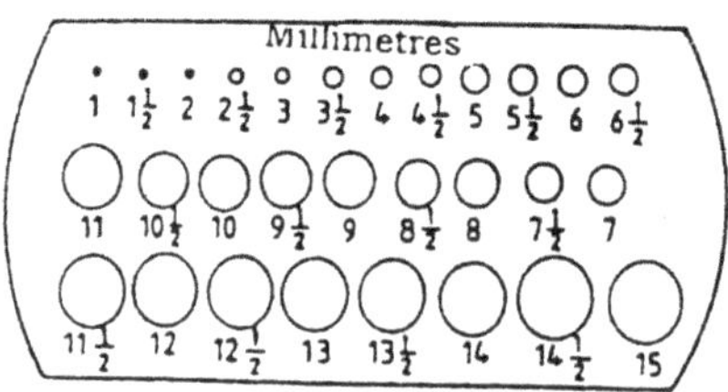

Drill guage

CUTTING TOOLS

Hacksaw

The hacksaw is used for cutting metal by hand. It consists of a frame which holds a thin blade, firmly in position. The blade has a number of cutting teeth. The number of teeth per 25mm of the blade length or teeth per inch (TPI) is selected on the basis of the work material and thickness (Table 1) being cut. Figure shows two types of hacksaw frames with a blade fixed.

The teeth of the hacksaw blade are staggered, as shown in Figure which is known as a "set of teeth". These make the slots wider than the blade thickness, preventing the blade from jamming.

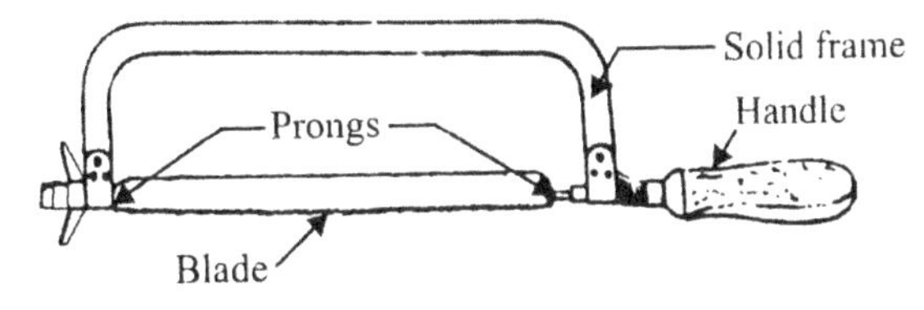

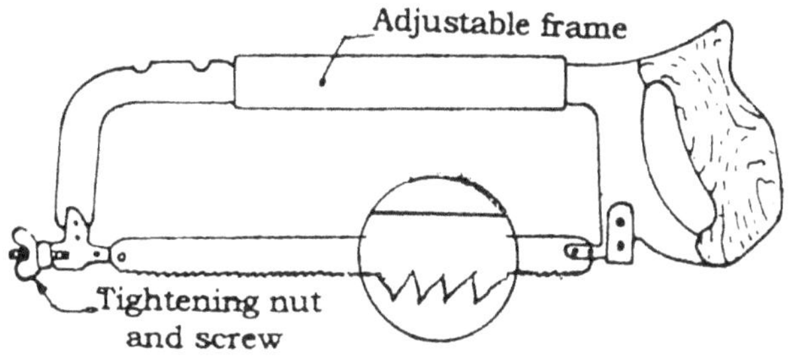

Hacksaw frame with blade

Set of teeth

Chisels

Chisels are used for removing surplus metal or for cutting thin sheets. These tools are made from 0.9% to 1.0% carbon steel of octagonal or hexagonal section. Chisels are annealed, hardened and tempered to produce a tough shank and a hard cutting edge. Annealing relieves the internal stresses in the metal. The cutting angle of the chisel for general purpose is 60 degrees.

A flat chisel is a common chisel used for chipping and cutting off thin sheet-metal

A cape chisel is narrow shaped tool. It is cased mostly for the clipping grooves and keyways.

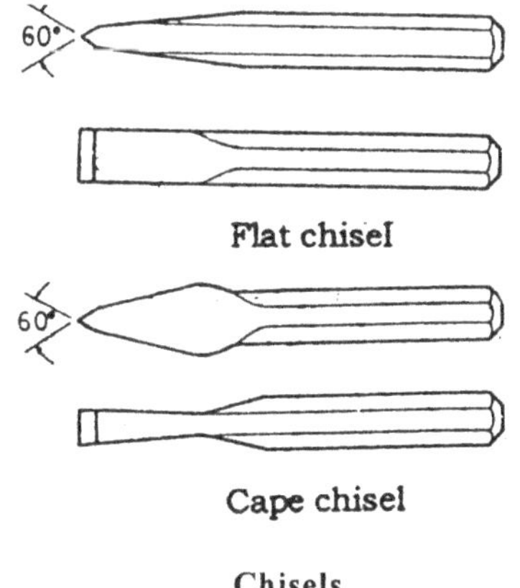

Chisels

Combination cutting plier

This is made of tool steel and is used for cutting as well as for gripping the work. The handles of the pliers used by electricians are insulated with PVC covering to protect from electric shocks.

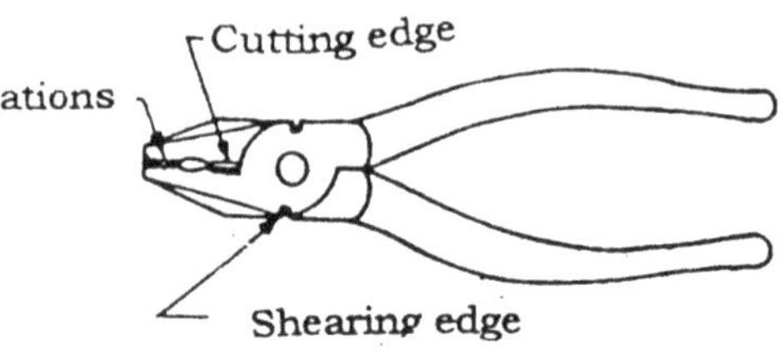

Combination cutting plier

Twist drill

Twist drills are used for making holes. These are made of high speed steel. Both straight and taper shank twist drills are used with machines. The following are the types, sizes and designations of twist drills:-

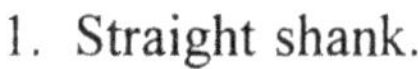

1. Straight shank.

Millimetres	from 0.4mm onwards
Inches	from 1/64" onwards
Letter drills	A to Z
Number drills	60 to 20

2. Taper shank

millimetres	3 to 100mm
Inches	1/8" to 4"

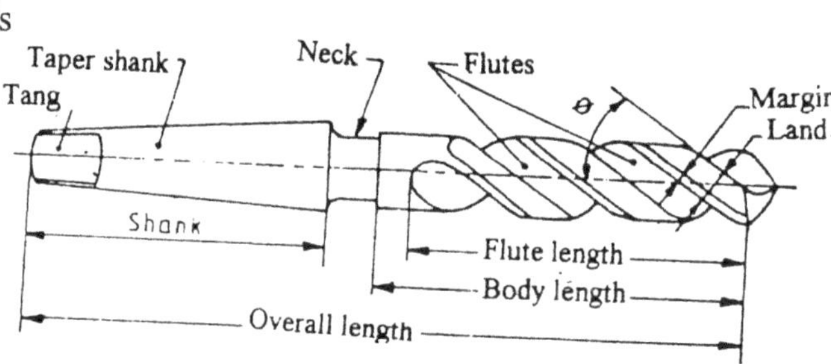

Twist drills

Taps and tap wrenches

A tap is a hardened steel tool, used for cutting internal threads after drilling a hole. Hand taps are usually supplied in sets of three for each diameter and thread pitch. Each set consists of a taper tap, intermediate tap and plug or bottom tap. The following are the stages involved in tapping operation:

1. Select the correct size tap, with the desired pitch. A thread is specified by its shape, size and pitch. Ex: M20 x 2.5 (nominal dia 20 mm, pitch 2.5 mm Metric thread).

2. Select the correct size tap drill, usually indicated on the tap.

3. Drill the hole.

Religion is a way of life. Silence is the song of the Soul.

4. Secure the tap in the tap wrench.

5. Insert the first or taper tap in the drilled hole and start turning clockwise, by applying downward pressure.

6. Check the alignment of the tap with the hole axis (verticality) with a try-square and correct it if necessary, by applying sidewise pressure while turning the tap.

7. Apply lubricant while tapping.

8. Turn the tap forward about half a turn and then back until chips break loose. Repeat the process until threading is completed with intermediate and bottom taps.

9. Remove them carefully. If it gets stuck, work it back and forth gently to loosen.

NOTE

1. It is good practice to drill a small countersunk, about the depth of one thread to ensure that a base is not thrown up while tapping the hole.

2. While tapping in a blind hole, remove the tap and clear the chips often so that the tap can reach the bottom of the hole.

Dies and Die-holders

Dies are cutting tools used for making external threads. Dies are made either solid or split type. They are fixed in a die holder for holding and adjusting the die gap. They are made of tool steel

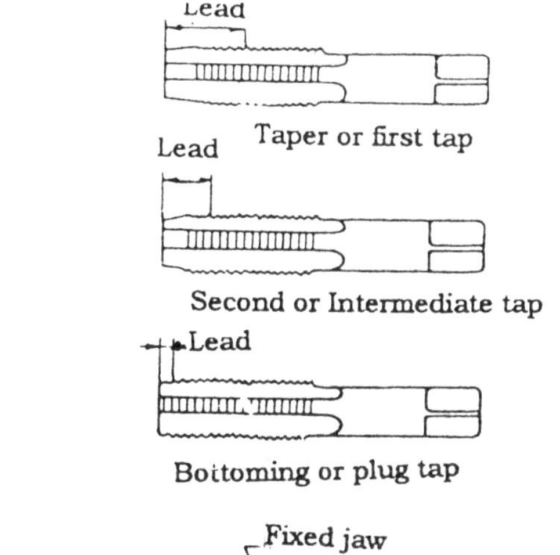

Taps and tap wrench

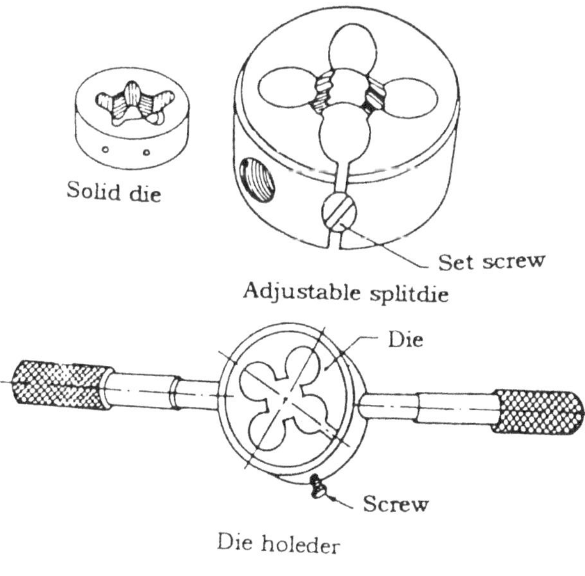

Dies and die handles

or high carbon steel. The following are the stages in producing external threads:

1.   Prepare the work with chamfer at its end.

2.   Select the correct size die.

3.   Position the die in the die holder.Tighten the set screw so that the die is held firmly in its place. In case of adjustable die, set the die to cut oversize threads first.

4.   Fasten the work firmly in a vice.

5.   Place the die over the chamfered end of the work and start cutting threads by turning it clockwise while applying downward pressure. Apply cutting fluid while threading in steel.

6.   Turn back the die for the chips to break loose. Continue until threading is completed.

7.   Check the threaded work to see if it fits the tapped hole or nut. If the fit is too tight, adjust the die for a slight deeper cut and complete the threading again.

NOTE : A tap is not adjustable, so it is better to tap first and then cut the external threads to fit the tapped hole.

Screw extractor (Ezyout)

Bolts, screws, studs, and other threaded parts may be sheared off, leaving a portion behind in the tapped hole. This portion can be removed by using a screw extractor. It is made of high-carbon steel, and has a tapered shape and left-hand threads for removing right-hand screws. A screw extractor of a size smaller than the broken screw is chosen from a range.

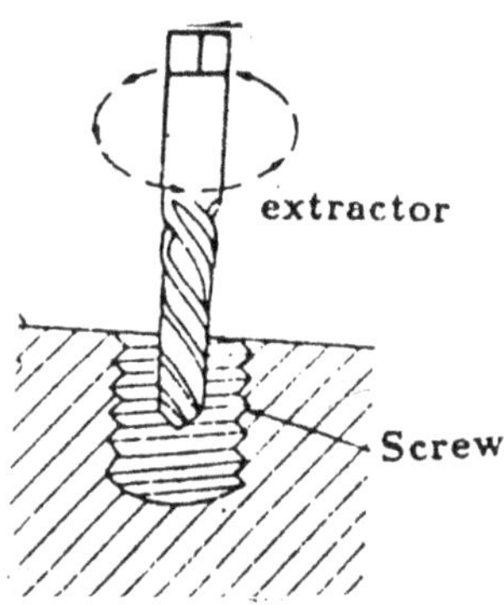

Screw extractor

Tap extractor

This tool is used to extract parts of taps that are broken in a hole. An extractor has prongs that fit into the flutes of a tap. The extractor is turned counter clockwise with a tap wrench to remove a broken right-hand tap.

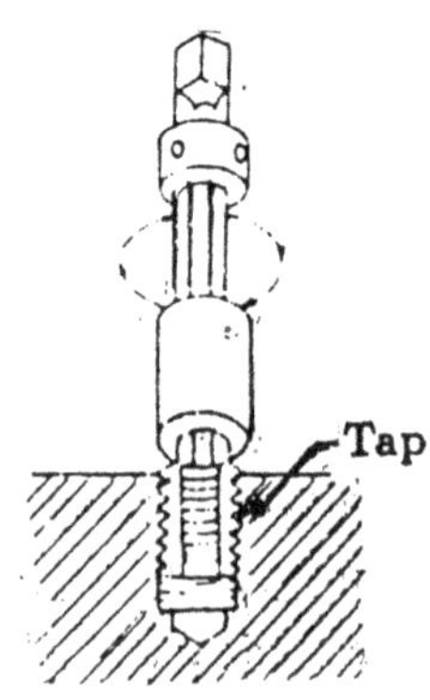

Tap extractor

FINISHING TOOLS

Files

Filing is one of the methods of removing small amounts of material from the surface of a metal part. A file is a hardened steel tool, having slant parallel rows of cutting edges or teeth on its surfaces. On the faces the teeth are usually diagonal to the edge. One end of the file is shaped to fit into a wooden handle. Figure shows the parts of a hand file.

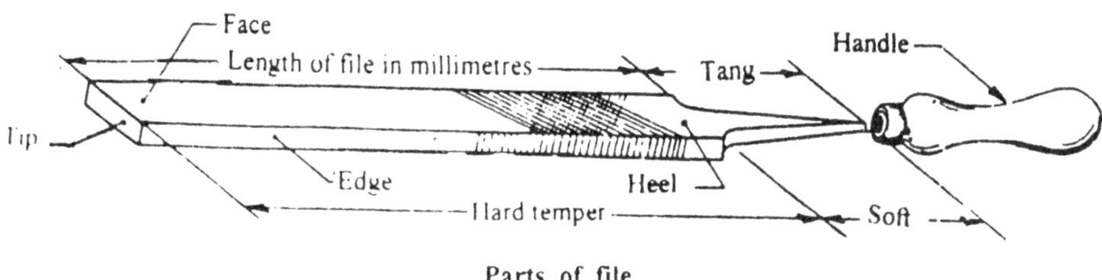

Parts of file

The hand file is parallel in width and tapering slightly in thickness, towards the top. It is provided with double cut teeth on the faces, single cut on one edge and no teeth on the other edge, which is known as the safe edge.

Types of files

Files are classified according to their shape, cutting teeth and pitch or grade of the teeth, Figure shows the various types of files in use based on their shape.

Type of file	Description and Use
1. Hand file	Rectangular in section and tapered in thickness but parallel in width. The faces carry double cut teeth and one of the edges single cut. The other edge, known as safe edge, does not have any teeth and hence this file in also known as safe edge file. It is useful in filing a surface which is at right angles to an already finished surface.
2. Flat file	It is rectangular in section and tapered for 1/3 length in width and thickness towards the point. The faces carry double cut teeth and the edges carry single cut teeth. It is a general purpose file.
3. Square file	It is square in section and carry double cut teeth on all the four faces. It is tapered for 1/3 of its length towards the point. Square files are used for filing corners and slots. It is also used to cut keyways and slots.

---

4. Three square file   It is of equilateral triangular in section and tapers towards the tip. The faces are double cut and the edges sharp. These files are used to file angular hole, and recesses. Used for sharpening wood saws.

5. Round file   It is tapered for 1/3 length with double cut on large coarse grades. Used for filing out round, elliptical and curved openings.

6. Half round file   The half round file has one flat and one curved side. The flat side is double cut and the curved side is single cut. It is not a semicircle but only about 1/3 of circle. Second cut and smooth grades are used. This is an extremely useful double purpose file for flat surfaces and for curved surfaces which are too large for the round file to be used.

7. Swiss or Needle files   150mm long with double cut teeth. Used for filing corners, grooves, narrow slots, etc.

Cut refers to 'single cut' and 'double cut' files. Single cut files have rows of teeth running in one direction,across their faces and double cut files have a second row of teeth cut

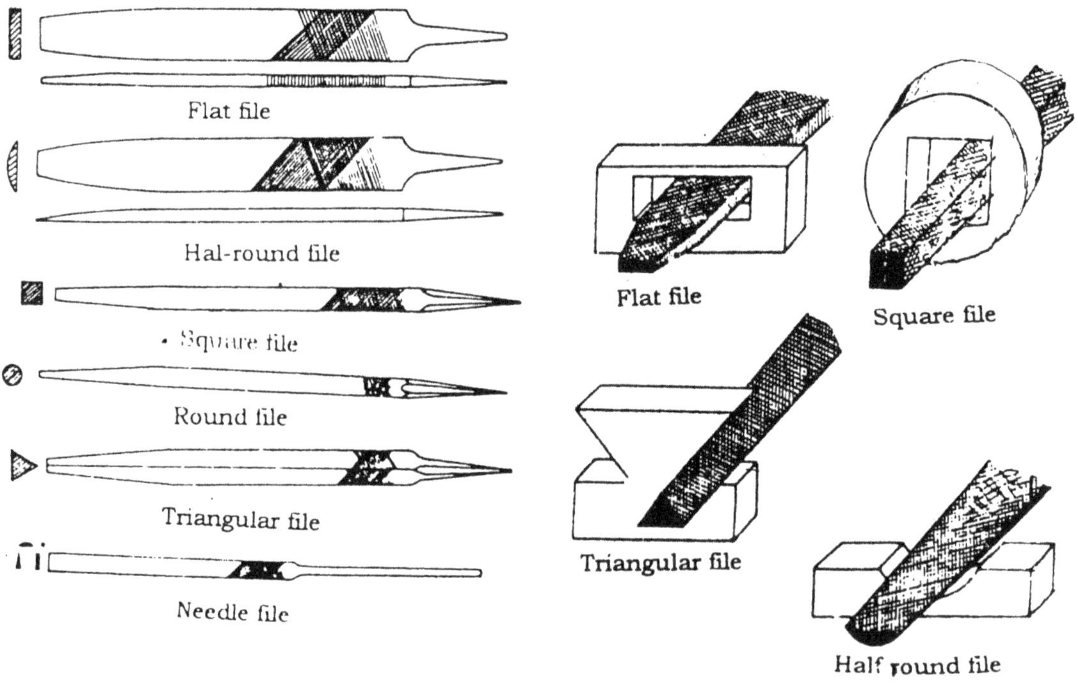

You are unique in this world. There is no one exactly like you.

diagonally to the first row as shown. Single cut files are used with light pressure to produce smooth finish. These are widly used for finishing on turning jobs. Classification of files based on the grade or the pitch of the teeth, is shown in figure.

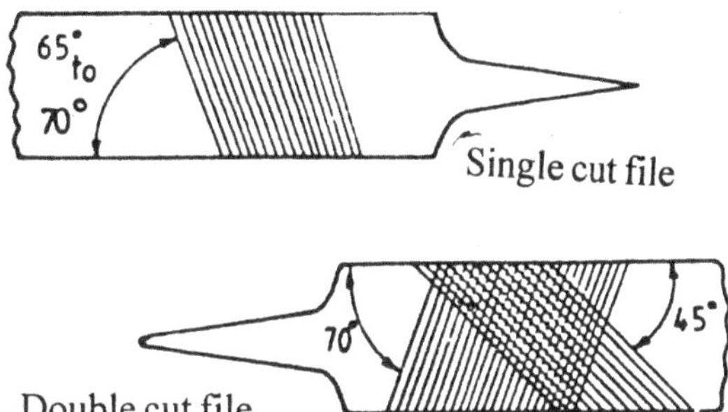

Cut of files

File card

It is a metal brush used for cleaning the files, to free them from filings, clogged inbetween the teeth.

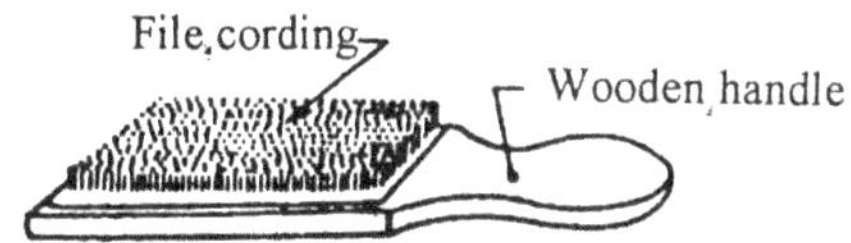

File card

MISCELLANEOUS TOOLS

Hammers

Hammers are named depending on their shape and material and specified by their weight. A ball peen hammer has a flat face which is used for general work and the ball end particularly used for revetting. They weigh from 200gm to 1.5kg.

The hammer consists of a hardened and tempered steel head varying in mass from 0.1 kg. To about 1kg, firmly fixed on a tough wooden handle.

Type	Form	No of teeth/cm
Rough		8
Bastard		12
Second cut		16
Smooth		20-24
Dead smooth		40

Grades of files

The flat striking surface is known as the face, and the opposite end is called the peen. The most commonly used is the ball-peen, which has a hemispherical end and is used for riveting over the ends of pins and rivets.

For use with soft metals such as aluminium or with finished components where the workpiece could be damaged if struck by an hammer, a range of hammers is available with soft faces, usually hide, copper, or a tough plastics such as nylon. The soft faces are usually in the form of replaceable inserts screwed into the end or forced into a recess in the face.

Always use a hammer which is heavy enough to deliver the required force but not too heavy to be tiring in use. The small masses, 0.1 kg to 0.2, are used for centre punching, while the 1 kg ones are used with large chisels or when driving large keys or collars on shafts. The length of handle is designed for the appropriate head mass, and the hammer should be gripped near the end of the handle to deliver the required blow. To be effective, a solid sharp blow should be delivered and this cannot be done if the handle is held too near the hammer head.

Always ensure that the hammer handle is sound and that the head is securely fixed.

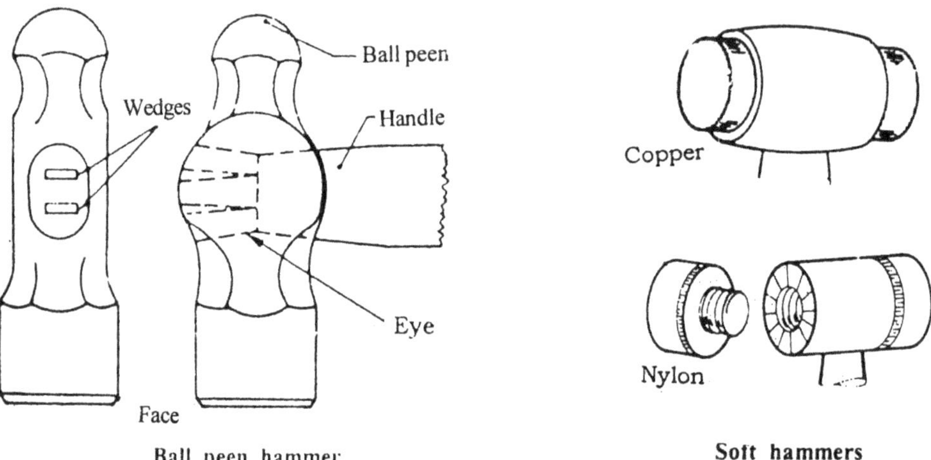

Ball peen hammer   Soft hammers

Spanners

A spanner or wrench is a tool for turning nuts and bolts. It is usually made of forged steel. There are many kinds of spanners as shown is figure. They are named according to the shape and application. The size of the spanner denotes the size of a bolt on which it can work.

Screw drivers

A screw driver is designed to turn screws. The blade is made of steel and is available in different lengths and diameters. The grinding of the tip of the blade to correct shape is very important. Screw driver is specified by the length of the steel rod. Screw drivers with small diameter rods are known as connectors. For better grip on screws which are small and at not easily accesable depths, philips (star) screw drivers are used. The end of the blade is fluted.

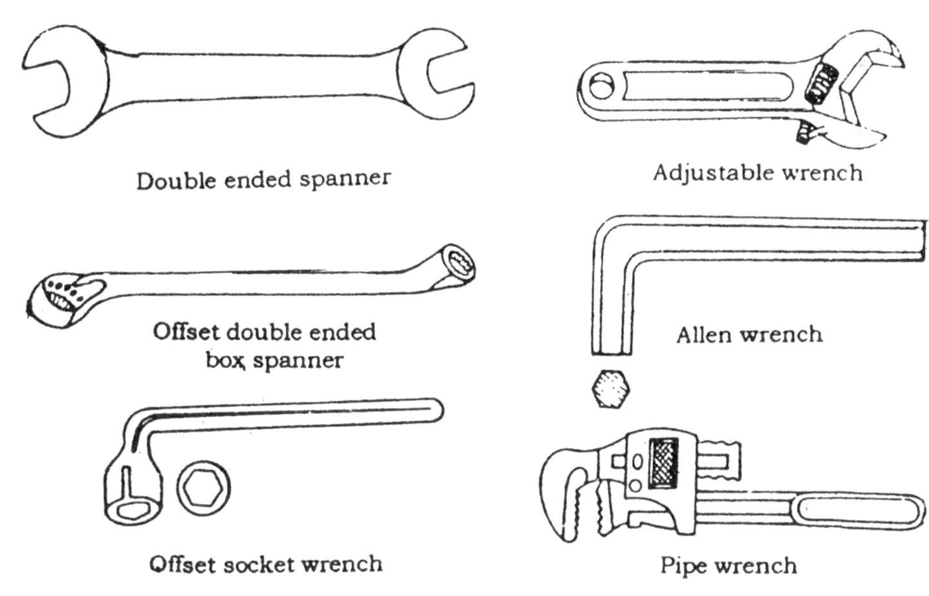

Spanners

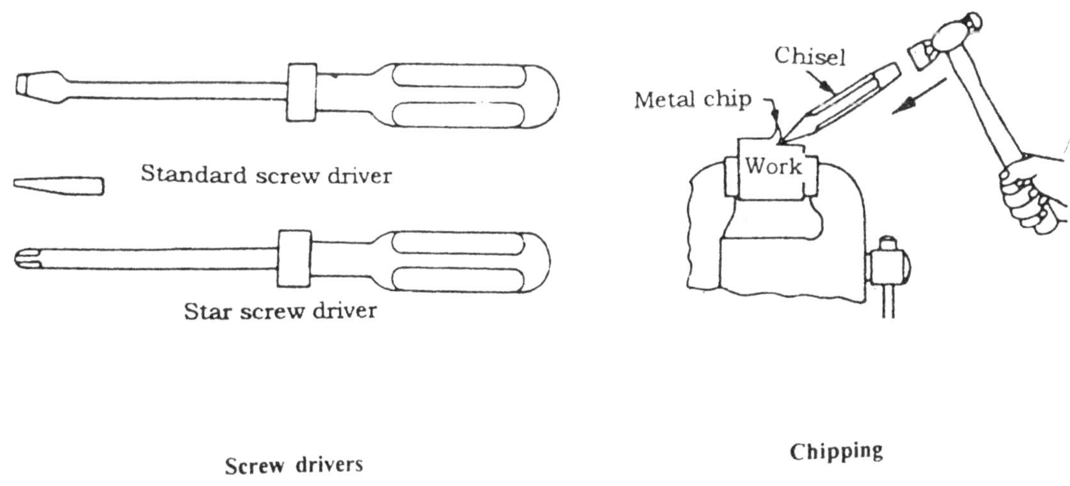

Screw drivers

Chipping

FITTING OPERATIONS

Chipping

Removing the metal with a chisel is called chipping and is normally used where machining is not possible. While chipping,safety goggles must be put on to protect eyes from the flying chips. To ensure safety of others, a chip guard is placed in position. Care should be taken to see that the chisel is free from mushroom head.

Filing

There are several methods of filing, each with a specific purpose. With reference to the figure, the followng may be noted:

1. **Holding the file:** For heavy work and to remove more metal, a high pressure is used. For light and fine work, a light pressure is applied.

2. **Filing internal curves:** A part of half round file only makes contact as shown during filing operation. Movement of the file is indicated by arrows.

3. **Cross filing.** It is the most common method of filing. Cross filing is carried out across two diagonals, to produce medium surface finish. It is used when large amounts of metal is to be removed. By cross filing 'rounding' the surface is reduced.

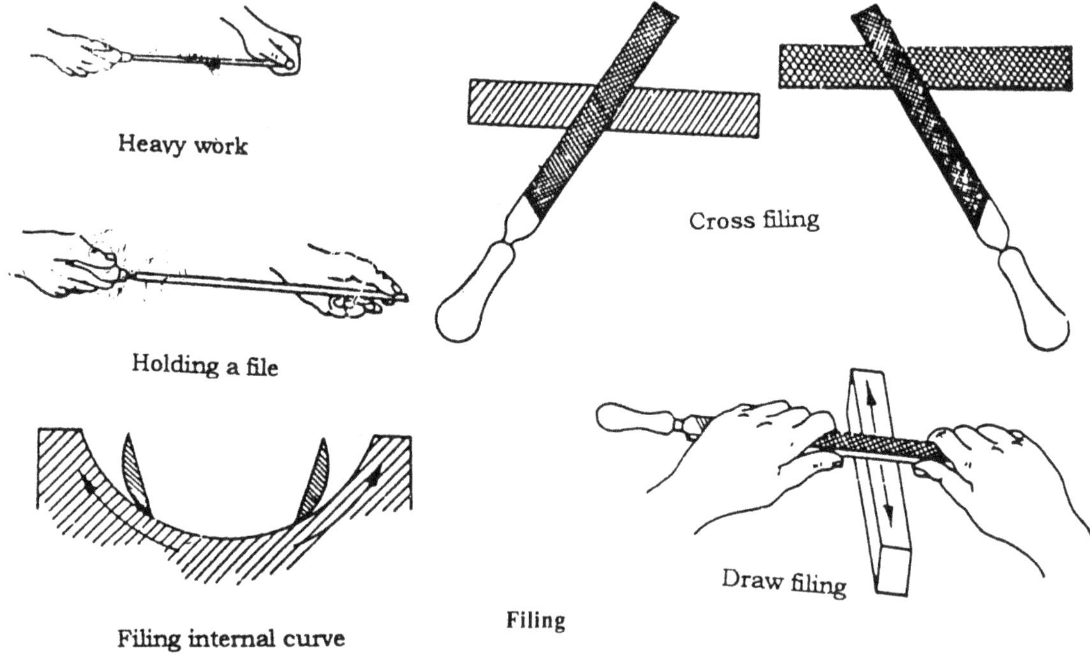

4. **Straight filing:** When a short length of workpiece is required to have a flat surface, straight filing is used. File marks made during cross filing may be removed, to produce a relatively smooth surface.

5. **Draw filing:** It is done to get a finely finished surface. It produces a smoother surface finish than straight filing. A smooth or dead smooth flat file is used for this.

Scrappers

These sharp edged tools are used to remove un-even spots on the surfaces. They are of different shapes.

Flat scraper

It is used for removing metal from flat surfaces. The blade must have a slight curvature at the cutting edge. The corners are rounded to help the user, scrape at the exact spots.

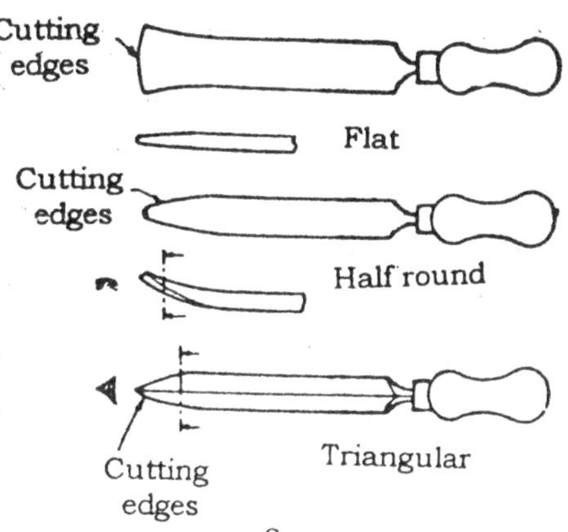

Half round bearing scraper

This is used for scraping curved and cylindrical surface – split bearings, big bush bearings etc.

Triangular scraper

This is used for scraping curved surfaces, holes and bores. Specification length. Example : 20mm, 30 mm etc.

Pinning of files

This is caused by soft metals, clogging the file teeth and scratching the surface of the work. The pins are removed with a file card. Pinning may be prevented by rubbing chalk into the teeth before filing.

Checking flatness and squareness

To check flatness, the try square is placed as shown in figure. Day light should not be seen between the bottom edge of the square and the

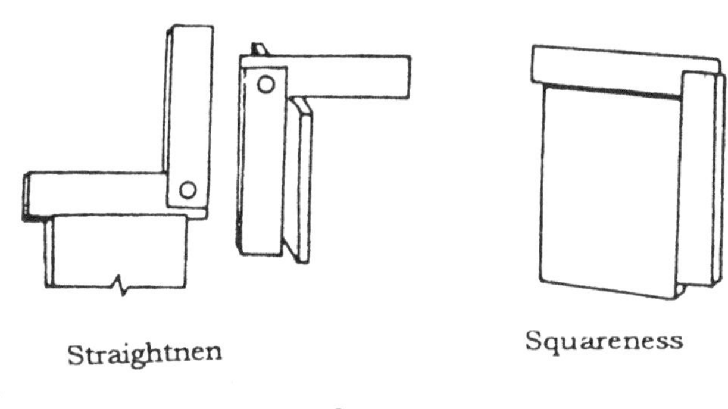

---

top surface of the work piece, when both are held against light. Similarly, the flatness across thickness of plate is tested as shown.

The squareness of one edge with respect to another reference edge is checked as shown in figure.

MARKING AND MEASURING

Accurate marking is the first step; and the methods and instruments used are common in all fitting works. Measurements are taken either from a finished edge or from a centre line.

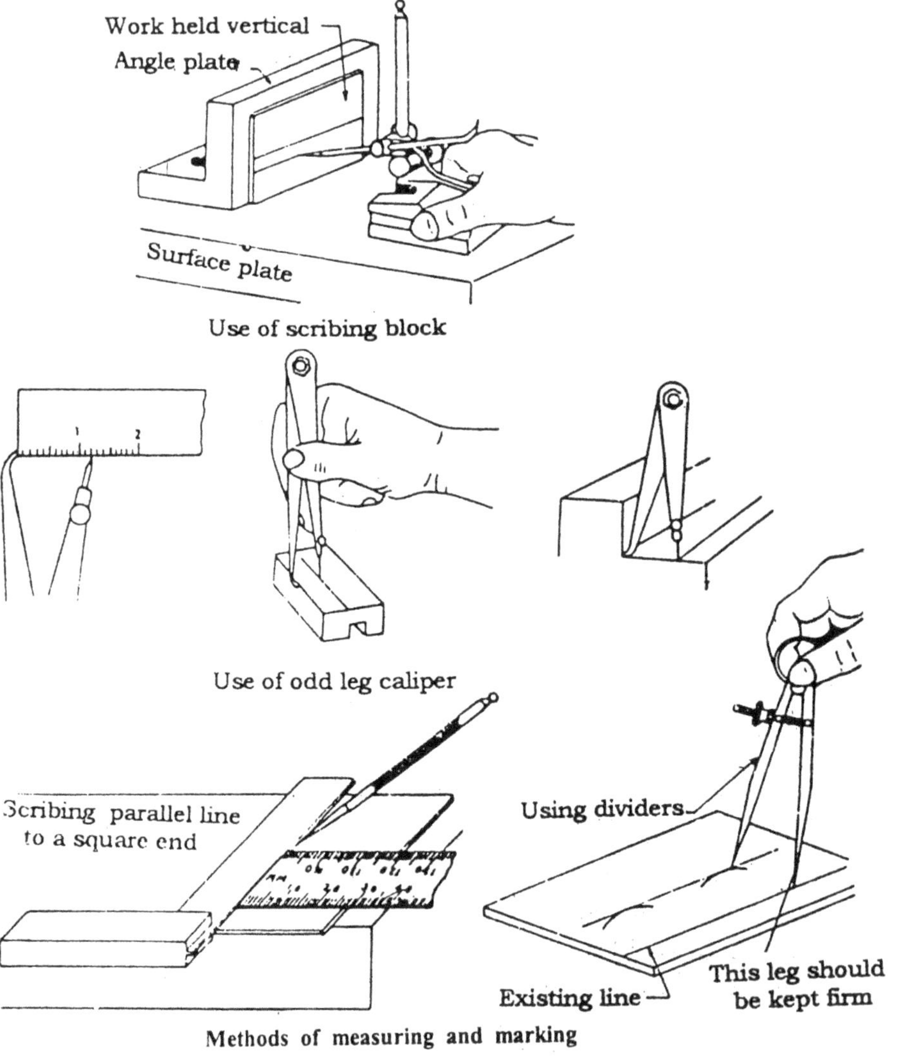

Methods of measuring and marking

Scriber lines on non-ferrous materials and oxide coated steels are readily visible but bright steel needs coating with copper sulphate solution or engineers blue (persian blue), for the visibility of the line.

Measuring and testing are continuous processes throughout the manufacturing, whether working with hand tools or machines. Degree of accuracy is specified on the drawing. The following are the measuring methods in the order of increasing accuracy:

(a) Direct measurement from a rule, (b) Caliper set to a rule, (c) Caliper set to a plug gauge, (d) Vernier calipers, (e) Micrometer, (f) Dial indicator

Figures, A to D show some methods of measuring and marking. Firm joint or spring calipers are used for transfering the dimensions, both external and internal, as shown in figure

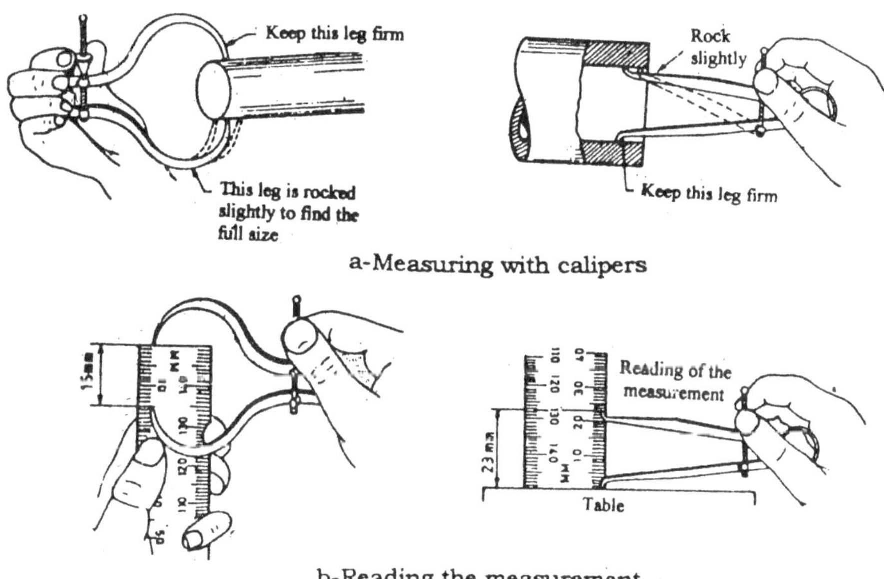

Measuring with calipers

Marking and Measuring Angles

Protractor

Protractor is used to measure angles. However, it is essential to position the protractor to the correct face of the component, as shwon in figure. Engineers protractor is marked in degrees and with care, reading to the nearest half degree may be made.

Vernier bevel protractor

A vernier protractor is used to measure angles upto an accuracy of 5 minutes. The vernier scale is divided into 24 parts, 12 on either side of the zero, each representing 5 minutes as shown in figure.

Priciples of Sawing

Hacksaw blades are specified by their total length, the width, thickness and the teeth provided per 25mm length called the pitch. Example: 300 x 12.7 x 0.65 x 18TPI. The correct choice of pitch should ensure that at least three teeth are in contact with the section under the saw. Blades should be inserted with the teeth pointing forward, as the saw cuts on the forward stroke only. Hacksaw blades are made in (i) All hard low alloy steel, (ii) High speed steel (HSS), and (iii) Flexible HSS. Little downward pressure is needed in sawing, as the teeth are designed to pull themselves into the work. About 40 strokes per minute is the correct sawing speed. HSS flexible blade manufactured by Bipico Eclipse India is ideal for students (beginners) as it cuts even hard steel easily and does not break at all. It has long life too for the extra cost paid.

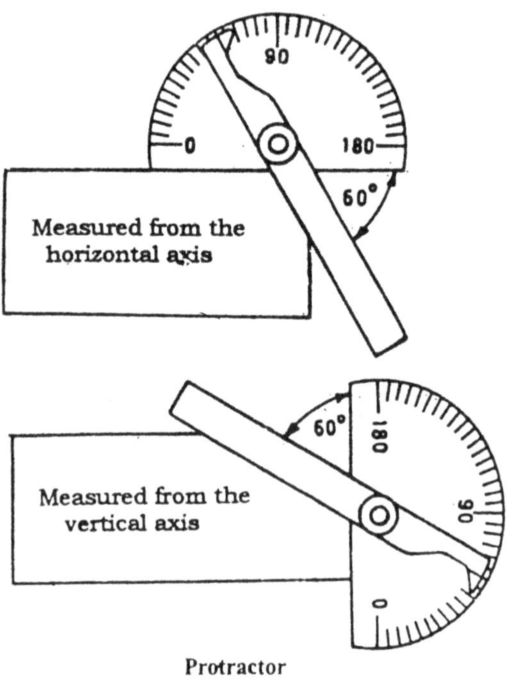

Protractor

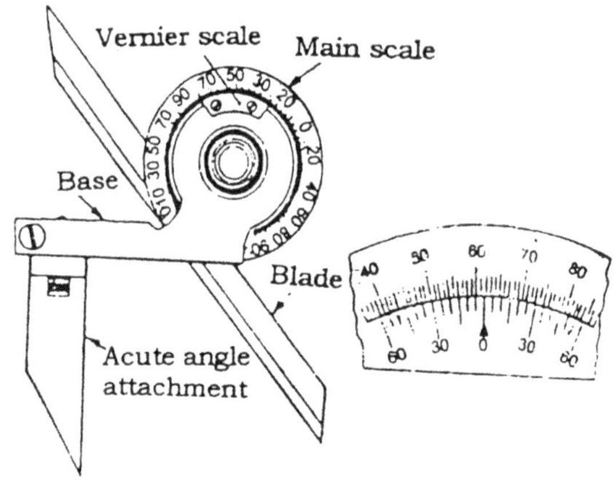

Vernier bevel protractor

Table 1 gives the application of the blades with respect to the pitch of the blade and material thickness. Table 2 shows the shapes or sections of some common raw materials available in the market.

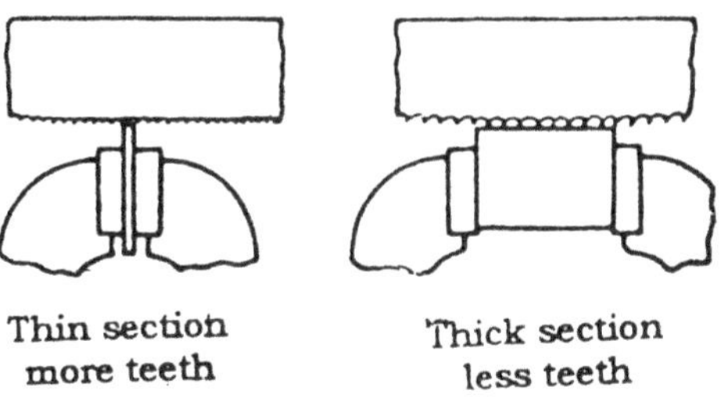

Table.1 Hacksaw blades application

Teeth per 25mm	Material Thickness, mm	Application
32	Upto 3	Thin sheets and tubes, hard and soft materials (thin sections)
24	3 to 6	Thicker sheets and tubes, hard and soft materials (thicker sections)
18	6 to 12	Heavier sections made of mild steel, cast iron, aluminium, brass, copper and bronze
14	12 to above	Soft materials of heavy sections, such as aluminium, brass, copper and bronze

SAFE AND CORRECT PRACTICES

The following are some of the safe and correct work practices in bench work and fitting shop:

1. Position the work piece area such that the cut to be made is close to the vice. This practice prevents springing, saw breakage and personal injury.

2. Use soft jaws when holding finished work surfaces in a bench vice.

3. Position the work in a vice so that it does not overhang into an aisle of other area where a person might accidentally brush against it.

4. Select the hacksaw blade pitch and set, most suitable for the material and the nature of the cutting operation.

5. Apply force only on the forward (cutting) stroke, relieve the force on the return stroke.

6. Start a new blade in another place when a blade breaks during a cut. This prevents binding and blade breakages.

Description	Section
Sheet upto approx .7 mm thick	 —
above 7 mm thick	▭
Bars – Round, Square, Octagonal	◯ ▢ ⬡
Flats	 ▭
Wires	 ○ ∘ ·
Cables	(section)
Angles (equal or enequal)	 L
Tees	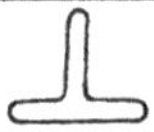 T
Channels	[ ]
H Section – beams	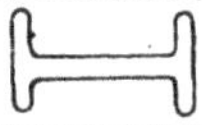 I
Extrusions (any type)	 (gear section)
Tubes (seamed or unseamed)	 ◎
Hollow , square or rectangular	 □

Forms of materials

7. Cut a small groove with a file in sharp corners, where a saw cut is to be started. The groove permits accurate positioning of the saw and also prevents stripping of the teeth.

8. For cutting thin metal strips, clamp them between two pieces of wood. Cutting through both the wood and the metal prevents the saw teeth from digging in and bending the metal.

9. Wear safety goggles or a protective shield when chipping and driving parts. Flying chips and other particles may cause eye injury.

10. Grind-off any mushroom that may form on the head of the chisel.

11. Use a file with a properly fitted, tight handle.

12. Examine the hammer each time before it is used. The handle must be securely wedged.

13. Select a screw driver with a tip which is only slightly smaller than the diameter of the screw head. The blade should fit snugly against the slot.

14. Select the type, shape and size of wrench opening, most suitable for the application. Position the wrench jaws as close to the work as possible, to prevent slipping.

15. Remove sharp projecting edges and burrs which produce inaccuracies in layout, measurement errors and improper fits.

MODEL RECORD SHEET

Aim : To make a step fitting as shown below :

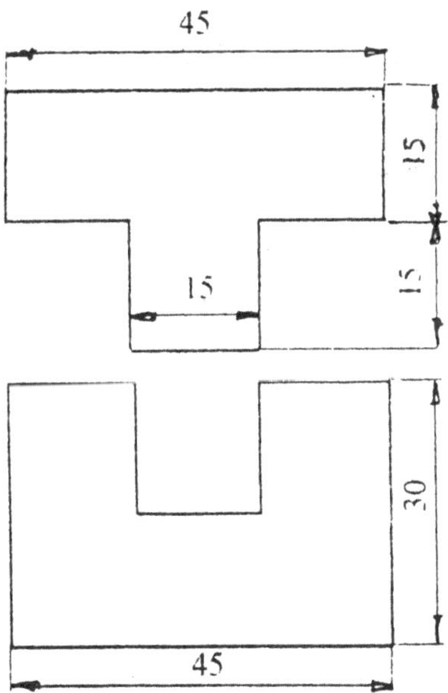

Material : MS Flat : 32mm x 50mm x 5mm thick. 2 Nos.

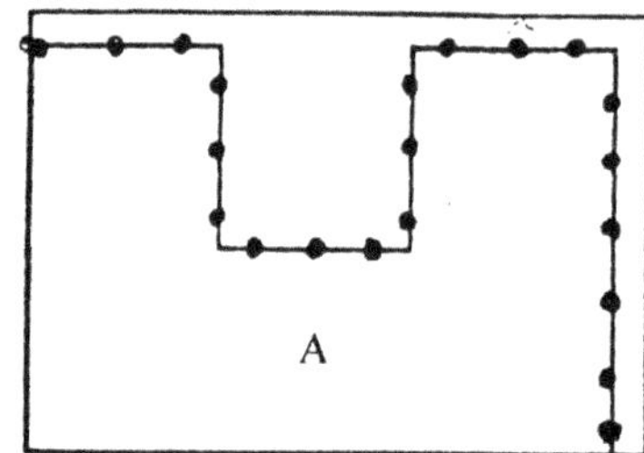 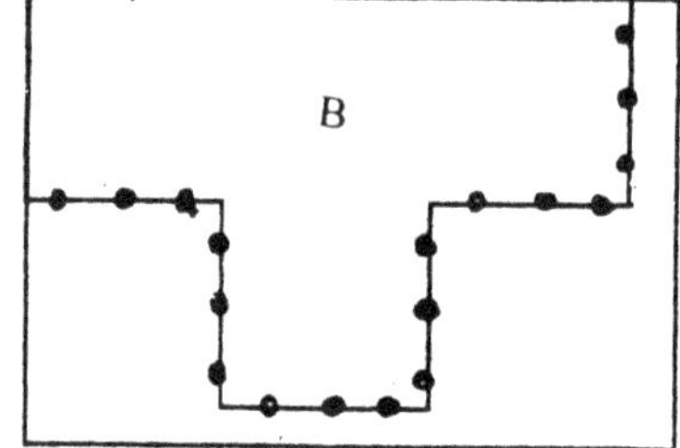

Marked and punched pieces

Tools required : 150mm steel rule, 150mm try square, 200gm ball peen hammer, dot punch, odd-leg caliper, 300mm hacksaw frame with 300 x 12.7 x 0 65 x 18 TPI hacksaw blade, 250mm rough and smooth flat files with safe edge, 10mm square file smooth, 150mm flat chisel.

Sequence of Operations

1.   The mild steel (MS) flats raw material given are checked for the dimensions.

2.   The 50 mm side is filed first with 250 mm rough file and then with smooth flat file.

3.   Its adjacent side of 35 mm is also filed with both rough and smooth files.

4.   The flatness and squareness of the edges to one another is checked with try square (see page 1.19)

5.   Both pieces are marked using angle plate, steel rule and scribing block (page 1.20) or odd leg caliper.

6.   Both pieces are punched with dot punch along the marking lines.

7.   The excess material over and above 30 and 45 mm is removed on both pieces using 300 mm hack saw.

8.   First the female piece (A) is cut by drilling holes within the marked lines and using chisel.

9.   The edges are now filed to get the flatness and finish.

10.   Final finishing is done using square files.

11.   The male piece (B) is cut along the dot punches. Then it is filed first with rough file and then with smooth file until it fits with the female groove.

12.   After fitting both pieces the top surfaces are filed to remove any burrs present.

Objective Questions

Fill in the blanks or answer the following in one word

1. The body of a bench vice is made of ____________ material.

2. ____________ are provided on the jaws of the bench vice to ensure good grip.

3. Jaw caps are made of ____________ material to protect finished surfaces

4. ____________ are used to support round jobs for layout and inspection

5. ____________ clamp is used to hold work.

6. For marking on heavy and big jobs ____________ is used

7. For marking or inspection of small jobs ____________ are used

8. ____________ is used to clamp jobs for marking or inspection

9. ____________ is used to scribe lines on the job

10. Try square is used to check ____________

11. Try square is specified by the ____________

12. Mitre of a combination set marks ____________ degrees on the work

13. ____________ is used to check squareness.

14. Odd-leg caliper is also called as ____________ and ____________

15. Drift punch i used to ____________

16. The point angle of a centre punch is ____________ degrees.

17. The point angle of a dot punch is ____________ degrees.

18. Before making fine adjustment of vernier ____________ clamp is locked.

19. Main scale readings of a micrometer are marked on the ____________ and Vernier readings on the ____________

20. Final pressure of the spindle of the micrometer on the job should be applied only by ____________ and not by thimble.

Answers

1. Cast Iron (C.I) 2. Serrations 3. Soft sheet metal 4. V-blocks 5. U-clamp 6. Marking table 7. Surface table 8. C-clamps 9. Scribing block 10. Flatness and sawareness 11. length of the blade 12. 90° 13. Square head 14. Jenny caliper 15. Allign holes 16. 90° 17. 60° 18. Screw on 19. Barrel and Thimble 20. Ratchet

---

Do only what you know to be right to gain real happiness.

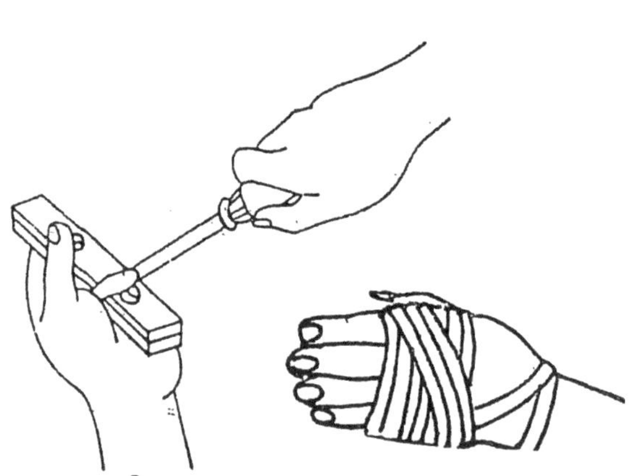

Using screw driver with work
held in hand may slip and cause
a serious wound.

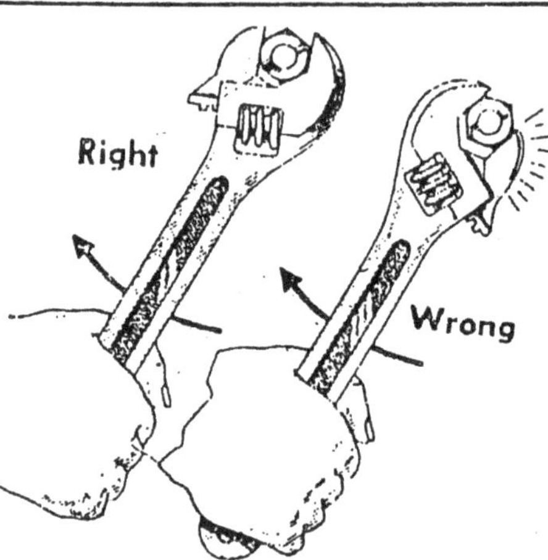

Right and wrong way of using
adjustable wrench.

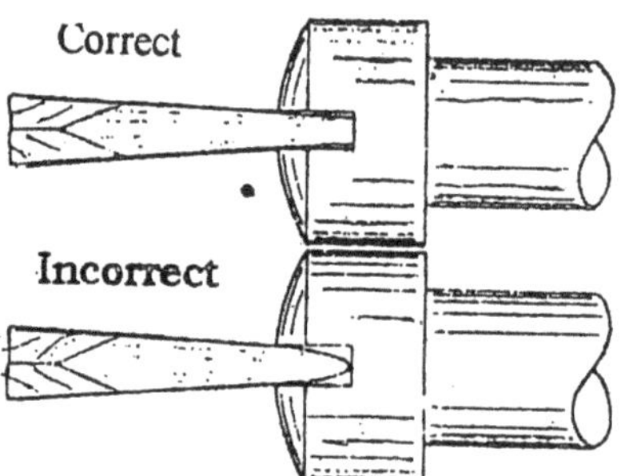

Grinding of screw driver blade

Use safety goggles
while grinding.

RIGHT AND WRONG WORKING

Right (✓) **Wrong (X)**

The following figures illustrate right and wrong ways of doing certain fitting operations and use of tools.

(a) With left foot set forward the whole of the body is in action and the filing or cutting stroke with pressure is done without much strain to hands or legs. In the second case with body movement the arm would tire soon.

(b) With filing diagonally across the work, smooth finish is obtained. Note that the file moves in the direction of the length of the file as shown. In the second case cut of the file teeth are produced on the work.

(c) Keep the work as low in the vise as possible. Work too high means lack of rigidity and too much vibration.

(d) Hold work within the width of the vise jaws, using the full grip of the vise. Avoid unnecessary overhang resulting in poor surface finish.

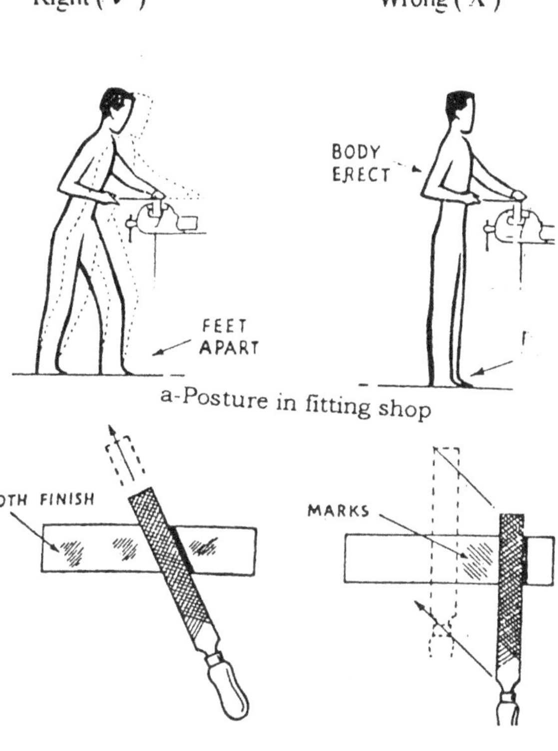

a-Posture in fitting shop

b-Filing flat surfaces

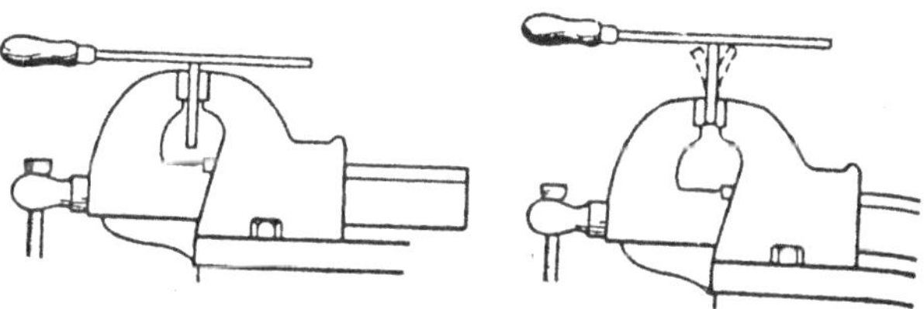
c-Holding work in vice

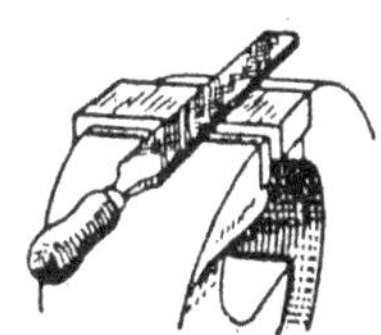

d-Filing with overhanging job

Knock the T off can't.

(e) **Work** across at an angle, left and right. It is a mistaken idea that filing along the length of the work produces a flatter surface.

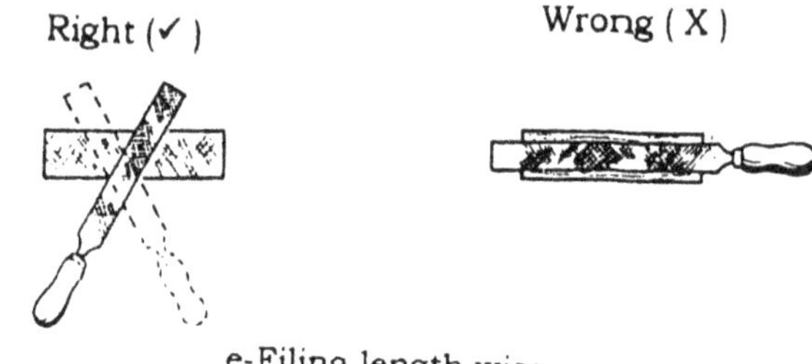

e-Filing length wise

(f) **Clamps** protect the surface of the finished job. without clamps jaw impressions are made on the finished surfaces.

f-Use of soft clamps

(g) **Body** action applies pressure on forward stroke and relief on return. Blade must be fitted to cut on forward stroke.

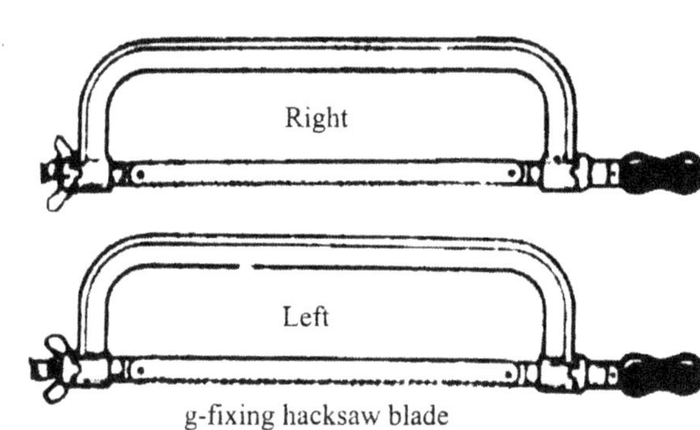

g-fixing hacksaw blade

(h) There must be more than one tooth in action.

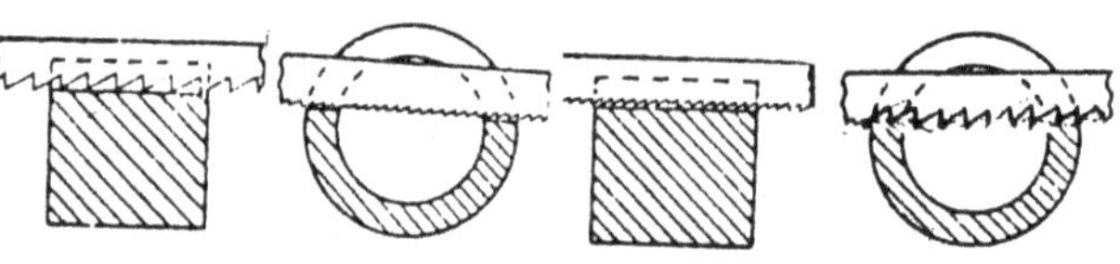

h-Selection of blade pitch

(i) Commence cutting with saw blade slightly inclined to the horizontal, picking up the line at far edge of work and proceed to horizontal position. In the second case it is difficult to pick up line accurately with blade engaging full width of the work. Blade too steeply inclined results in broken teeth.

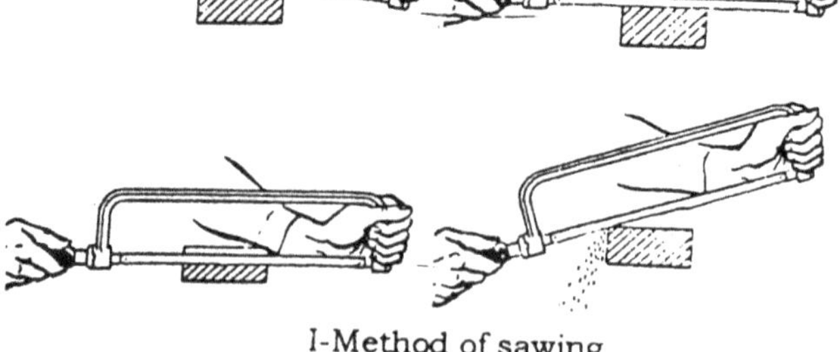

I-Method of sawing

Exercise - 1.1

Aim : To make a straight fitting.	Ex. No. :
	Date :

Sketch :

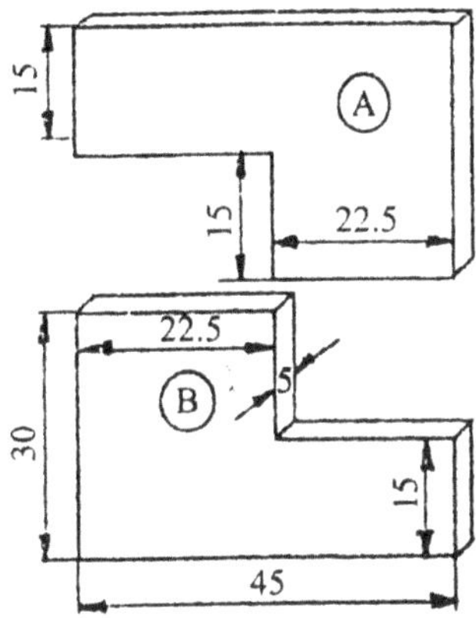

Material : MS flat

size 50 × 32 × 5 mm : 2 Nos.

List of tools used :

Sequence of operations :

Safety precautions :

Staff signature

Exercise - 1.2

Aim :	Ex. No. :
	Date :

Sketch :

Material :

List of tools used :

Sequence of operations :

Safety precautions :

Staff signature

Exercise - 1.3

Aim :	Ex. No. :
	Date :

Sketch :

Material :

List of tools used :

Sequence of operations :

Safety precautions :

Staff signature

Exercise - 1.4

Aim :	Ex. No. :
	Date :

Sketch :

Material :

List of tools used :

Sequence of operations :

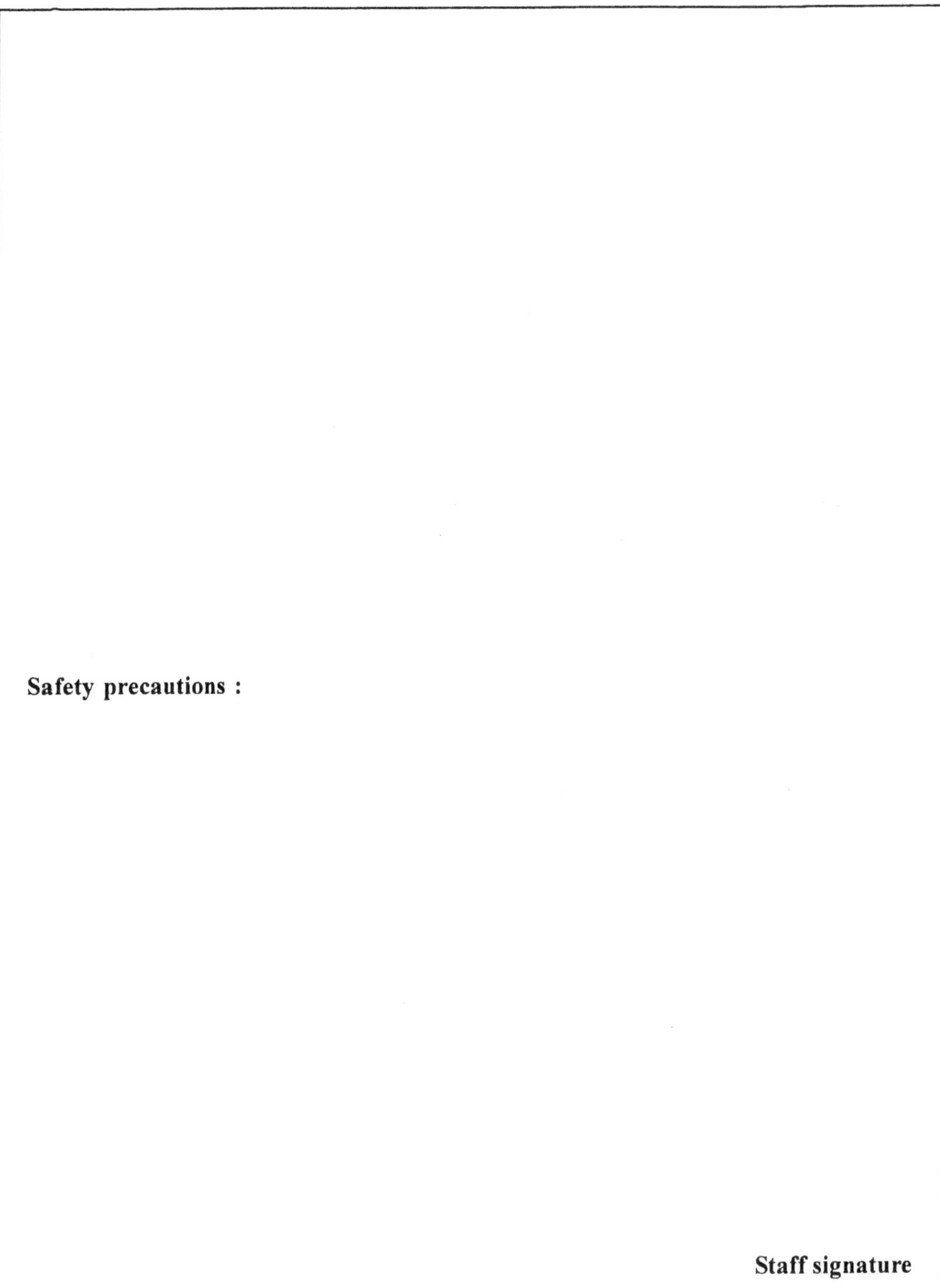

Safety precautions :

Staff signature

CHAPTER - 2

CARPENTRY SHOP

INTRODUCTION

Wood work or carpentry deals with making joints for a vareity of applications Viz Door frames, Window frames, Cabinet making, Furniture, Packing etc.

TIMBER

Timber is the name given to the wood obtained from well grown trees. The trees are cut, sawn into various sizes. The word 'grain' as applied to wood, refers to the appearance or pattern of the wood on the cut surfaces. The grains of the wood are of fibrous structure and to make them strong, the timber must be cut in such a way that the grains run parallel to the length. The common defects in timber are knots, cracks, wet rot, dry rot etc.

CLASSIFICATION OF TIMBER

According to the manner of growth the timber trees can be broadly classified as:

1. Exogenous or Outward growing

2. Endogenous or Inward growing.

Exogenous trees grow from the centre adding concentric layers of fresh wood every year, known as annual rings. These variety of trees yield timber suitable for buildings and engineering uses. The exogenous trees are further classified as:

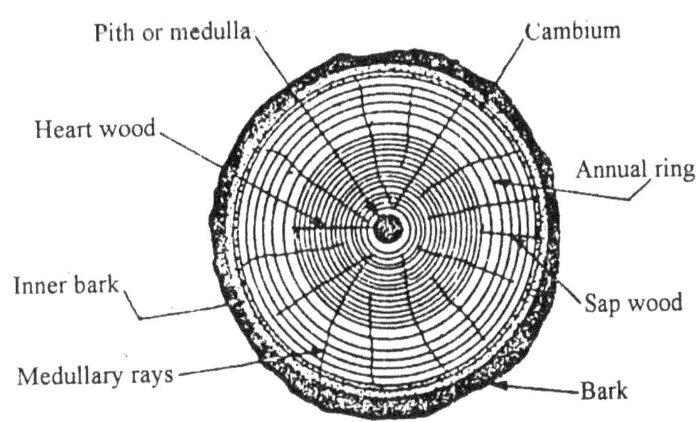

Cross-section of an exogenous tree

1. Conifers or evergreen trees.

2. Deciduous or broad leaf trees.

The conifers give soft wood and the deciduous class hard wood. Common examples of hard wood are Sal, Teak, Rose wood, Shisham, Oak, Beach, Ash, Ebony, Mango, Neem and Babid (Babool). Soft woods are Pine, Deodar, Kail, Chir, Walnut, Semal, Toon etc.

The other trees which grow inwards, ie. every fresh layer of sap wood is added inside instead of outside. These are known as endogenous and their common examples are Cane, Bamboo, Palm, Coconut etc.

The cross-section of an exogenous tree is shown in the figure.

Soft and hard wood

Softwood is obtained from trees having needle shaped leaves or conifers grown in colder part of the world and hard wood from those having broad leaves or deciduous trees in hot climate. The main characteristics of these two types of wood are given below:

Soft wood	**Hard wood**
1. It is light in colour.	It is dark is colour.
2. It is light in weight.	It is heavier.
3. It is easier to be worked on.	It is difficult to work on.
4. Gets splitted quickly.	Does not split easily.
5. Can catch fire soon.	More resistant to heat.
6. Relatively weaker and less durable.	Stronger and more durable, used for structural work.
7. Has a good tensile strength but is weak across the fibres.	Has resistance to both tensile and shear forces.
8. The annual rings are quite distinct	Not distinct in it.
9. Carries straight fibres and fine texture.	Fibres are compact.
10. It is a resinous wood having fragrant smell.	Non-resinous containing a fairly good amount of acid. Used for structural work.

Timber Sizes

Timber sold in the market is in various sizes and shapes. The following are the common shapes and sizes.

a) **Log**   - The trunk of the tree which is free from branches.

b) **Balk**   - The log, sawn to have roughly square cross section.

c) **Post**   - A timber piece, round or square in cross section, having its diameter or side from 175 to 300 mm.

d) **Plank**   - A sawn timber piece, with more than 275 mm in width, 50 to 150mm in thickness and 2.5 to 6.5 meters in length.

e) **Board**   - A sawn timber piece, below 175mm in width and 30 to 50mm in thickness.

f) **Reapers**- Sawn timber pieces of assorted and non-standard sizes, which do not confirm to the above shapes and sizes.

VENEERS

Veneers are thin layers of wood, made in different thicknesses, distinguished for their fine appearance and texture, lightness, strength and durability Veneered wood have less tendency to warp and shrink and their surfaces can be given an attractive and rich grained surface when compared to solid stock wood.

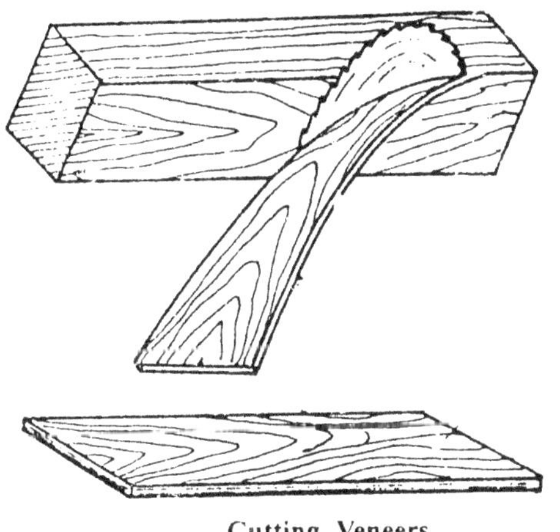

Cutting Veneers

The wood commenly used for veneering are Teak, Walnut, and Rose-wood. Logs of wood are converted into veneers by slicing as shown in figure. Thickness of these veneered sheets varies from 0.5mm to 6mm. They are glued to inferior timber surfaces to improve the appearance.

Veneers are used in the manufacture of plywood, laminated sheets. Veneers used for making plywoods are known on plies.

Auxiliary materials used in Carpentry

Nails : The nails used in wood work are either made out of drawn wires of low carbon steel or cut out of thin malleable steel rods. Those made from drawn wires are called wire nails and those made from rods as cut nails.

Wire nails are used for light and medium work and cut nails for heavy works. Nails used for fine and delicate work are called panel pins.

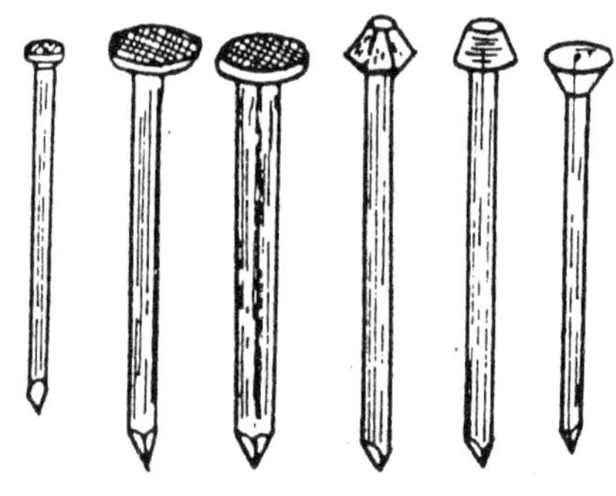

Various types of nails

Nails are mainly used for reinforcing glued joints and fastening different parts. Their size in designated by overall length and gauge (diameter) number and commercially sold by weight. The common shapes of nails are shown in figure.

Wood screws

These are made from bright drawn wires. Wood screws are specified by their length and wire gauge number.

Ply wood

Plywood is made of three or more layers. The central layer called core is usually thicker and relatively inferior wood than the face veneers. The veneers glued at the top and bottom are known as face-plies. The surface grains of adjacent layers are kept at right angles to each other. This arrangement prevents the ply wood from warping and shrinking. The total number of plies in ply wood is always odd number ie 3,5,7,9 etc. varying in thickness.

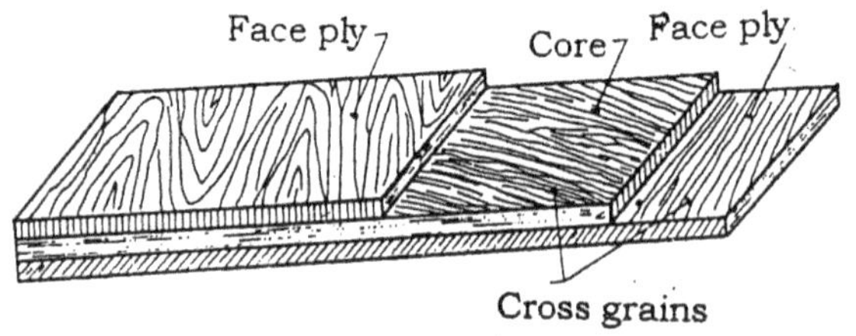

3 plywood

5 plywood

The plies may be cold pressed or hot pressed after applying glue to join them. The pressure and temper applied for hot pressing are about 7 to 14 kg/cm² and 150°C respectively.

Commercial ply woods

Commercial names in ply wood include **kit-ply** and **dura-tuff** which are hot pressed and water proof. Wood chips when glued and pressed result in particle boards or chip boards in varying thicknesses. These are given outside appearance by covering with a layer of lamination. **Novapan** is one such product having good surface finish. These particle bond sheets can not take compression and shear loads. Hence they are mostly used for partitions and panels for rich appearance at economical cost.

Block Board

These boards consist of solid Wood strips in the core and veneers on either side. The solid wood strips upto 20 mm wide are glued side by side as shown.

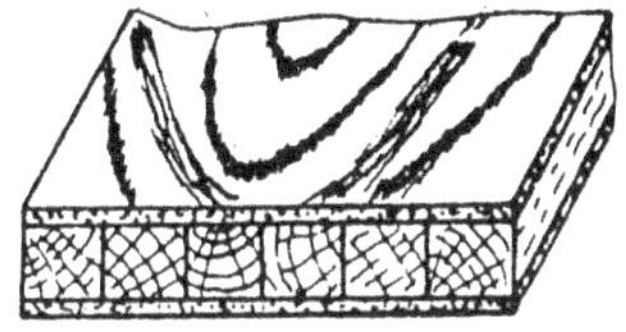

Block board

WOOD JOINTS

There are many kinds of joints used to connect wood stock. Each joint has a definite use and requires marking, cutting and joining together. The strength of the joint depends on the

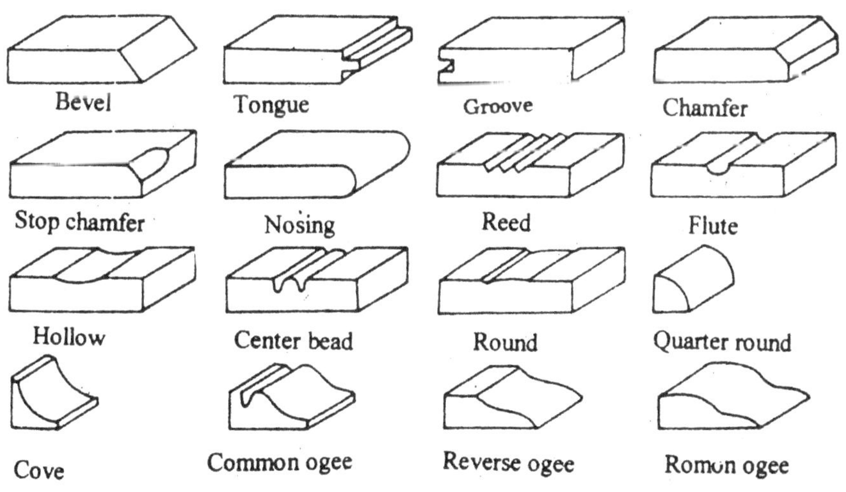

Common wood cuts

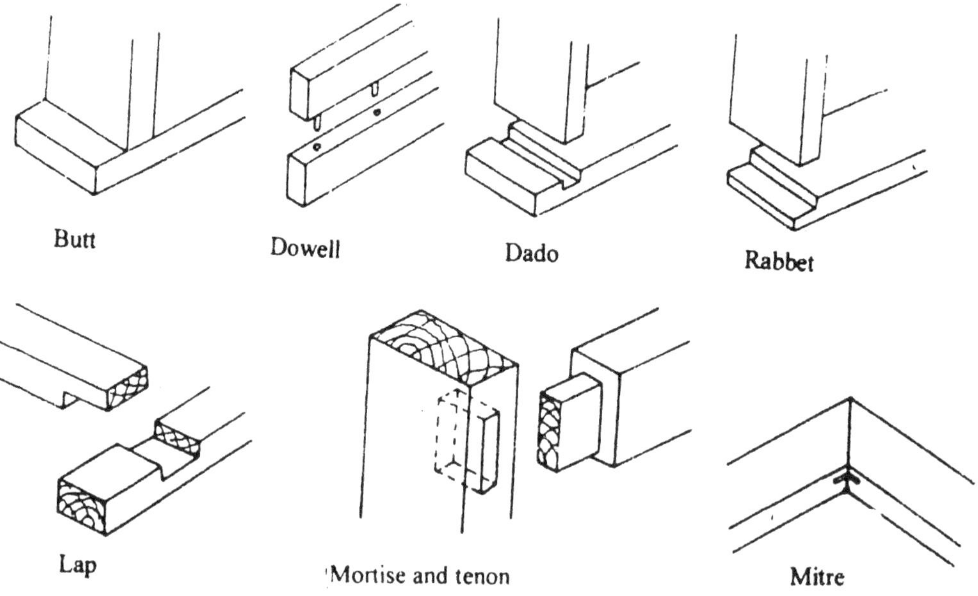

Common wood joints

amount of contact area. The joints are strengthened by using nails, screws and dowels wherever necessary. Figure shows some commnonly used wood cuts and joints. The nails are specified by their length and wire diameter which is indicated by wire gauge number. Similarly the wood screws are specified by their lengths and gauge numbers. Ex: wood screw 35 x 8 will be bigger in diameter than 35 x 12.

Lap Joints

In lap joints, an equal amount of wood is removed from each piece, as shown in figure Lap joints are easy to mark, using a try-square and a marking gauge. If the joint is found to be tight, it is better to reduce the width of the mating pieces instead of trimming the shoulder of the joint.

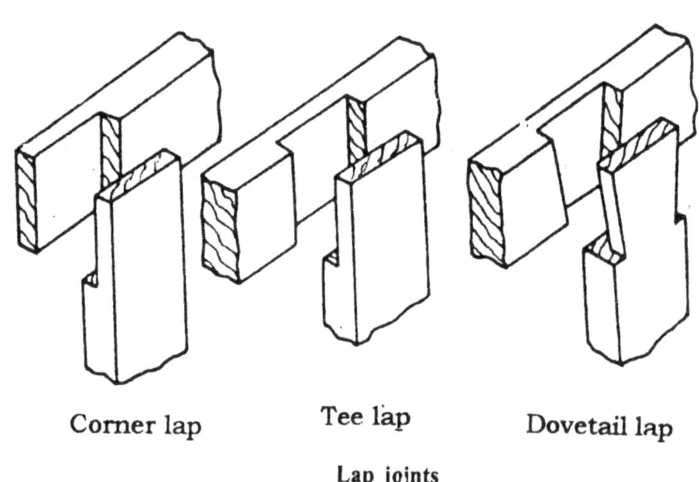

Corner lap   **Tee lap**   **Dovetail lap**

Lap joints

Mortise and Tenon Joints

These joints are used for the construction of quality furniture.

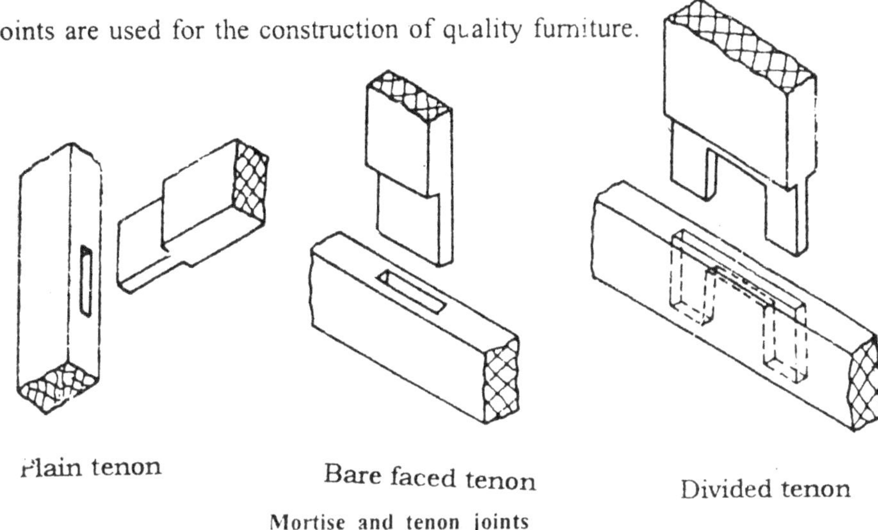

Plain tenon   Bare faced tenon   Divided tenon

Mortise and tenon joints

The joint results in a strong one and requires considerable skill to make it. The following are the stages involved in the work :

1. Mark the mortise and tenon layouts.
2. Cut the mortise first by drilling a series of holes within the layout lines, chiselling out the waste stock and trimming the corners and sides.
3. Prepare the tenon by cutting and chiselling.
4. Check the tenon size against the mortise that has been prepared and adjust it if necessary.

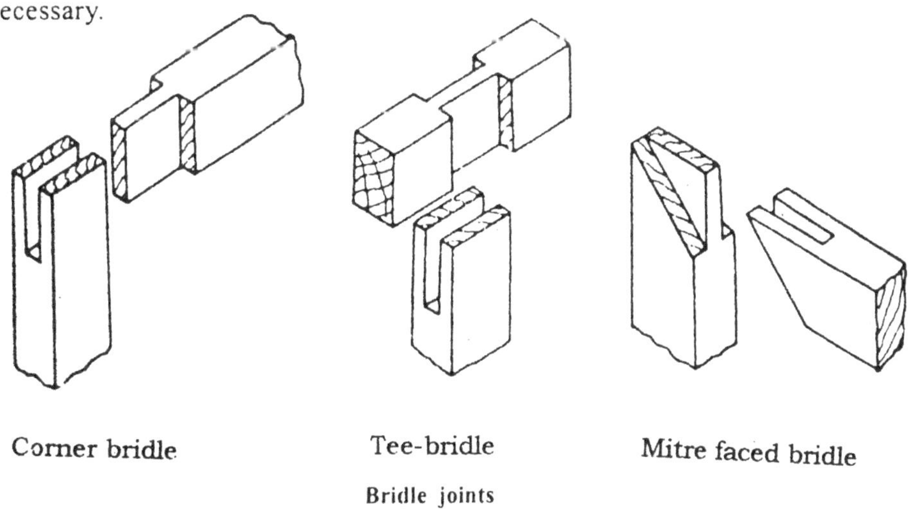

Corner bridle   Tee-bridle   Mitre faced bridle

Bridle joints

Bridle joints

These are the reverse of mortise and tenon joints in form. The marking out of the joint is the same as for mortise and tenon joint. These joints are used when the members are of square section and unsuitable for mortise and tenon joint.

WORK HOLDING TOOLS

Carpenters Vice

Carpenters bench vice in used to hold work prices for planing, cutting and chiseling, The Jaws of the vise are lined with hard wood faces.

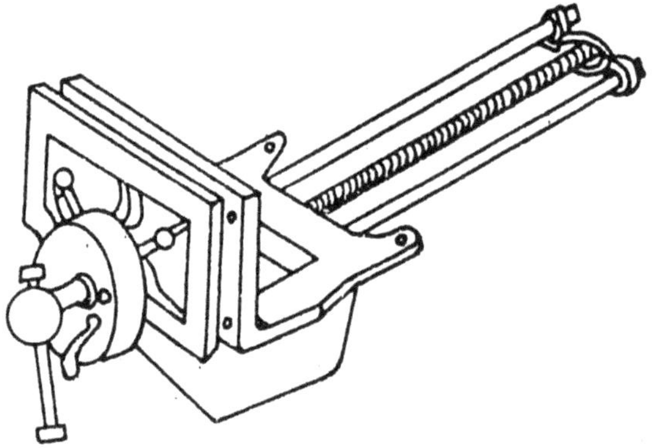

Carpenter's vice

Bar-Cramp

It is made of steel base Tee-section, with malleable iron fittings and a steel screw. It is used for holding wide works such as frames for tight gripping before nailing and putting dowels.

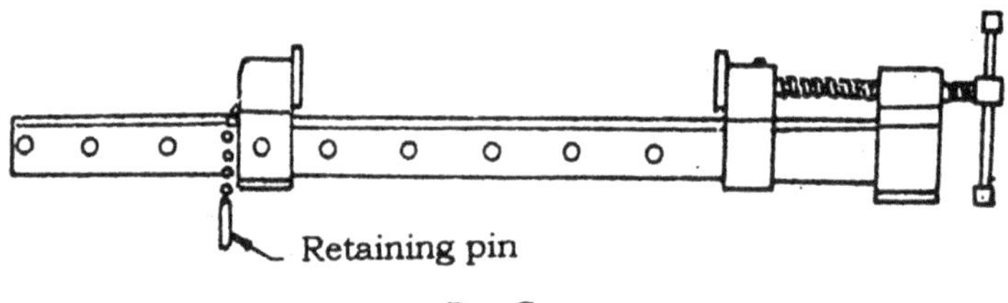

Bar-Cramp

MARKING AND MEASURING GAUGE

Marking Gauge

It is a tool used to mark lines parallel to the edge of a wooden piece. It consists of a square wooden stem with a sliding wooden stock (head) on it. A marking pin, made of steel is fitted on the stem. The stock is set at any desired distance from the marking point and fixed in position by a screw. It must be ensured that the marking pin projects through the stem about 3mm and the end is sharp enough to make a very fine line. A mortise gauge consists of two pins. In this, it is possible to adjust the distance between the pins, to draw two parallel lines on the stock.

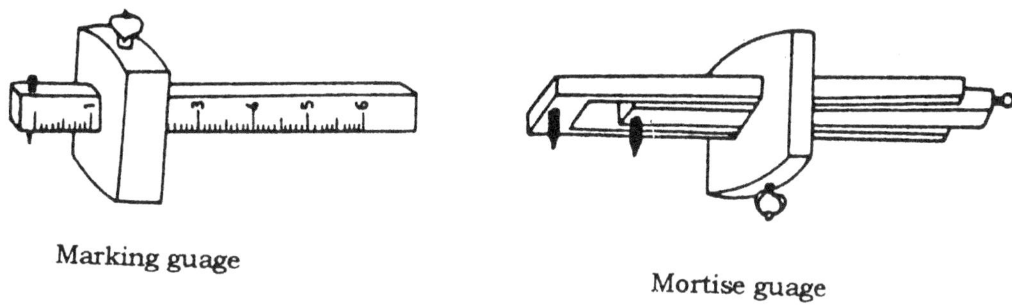

Marking guage

Mortise guage

Marking and mortise gauges

Try-square

It is used for marking and testing the squareness of planed surfaces. It consists of a steel blade, fitted in a cast iron stock. It is also used for checking the planed surfaces for flatness.

The size of a try-square varies from 150 to 300mm, according to the length of the blade. It is less accurate when compared to the try-square used in fitting shop.

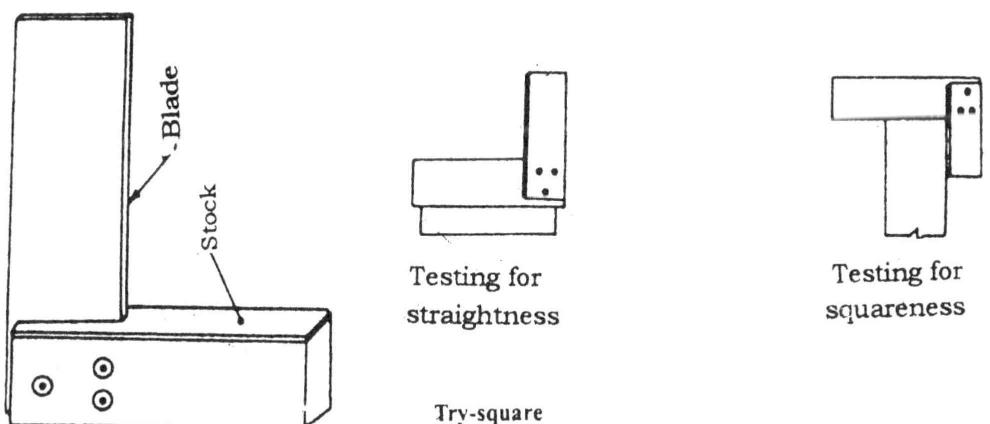

Try-square

Compass and divider

Compass and divider are used for marking arcs and circles on the planed surfaces of the wood.

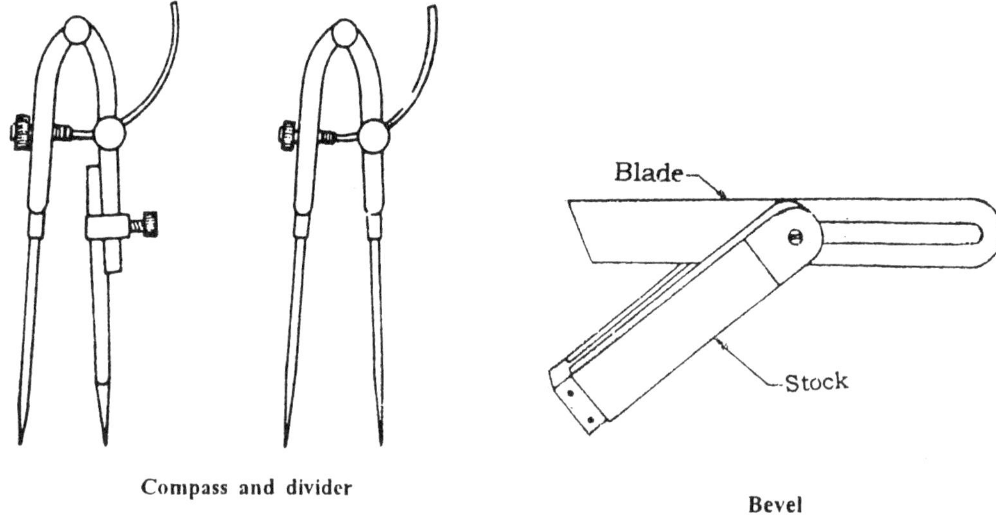

Compass and divider

Bevel

Bevel

It is used for laying out and checking angles. The blade of the bevel is adjustable and may be held in place by a thumb screw. After it is set to the desired angle, it can be used in the same way as a try-square.

PLANING TOOLS

In general, planes are used to produce flat surfaces on wood. The cutting blade used in a plane is very similar to a chisel. The blade of a plane is fitted in a wood or metallic block at an angle.

Jack plane which is about 350mm long is used for general planing. A smooth plane that is about 200 to 250mm long is used for smoothing the stock. Being short it can follow even the slight depressions in the stock, better than the jack plane.

A rebate plane is used for making a rebate. A rebate is a recess along the edge of a piece of wood which is generally used for positioning glass in frames and doors.

A plough plane is used to cut grooves, which are used to fix panels in a door. Figure shows the various types of planes described above.

Any fool can criticize, condemn and complain - and most fools do.

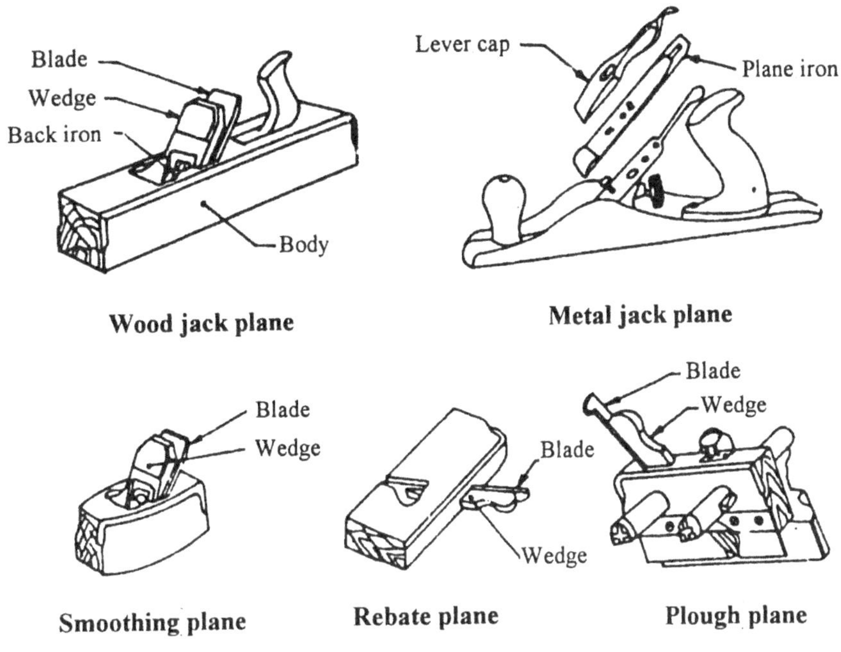

Types of planes

Cutting Tools

Hand saw or cross-cut saw

It is used to cut across the grains of the stock. The teeth are so set that the saw kerf will be wider than the blade thickness.

Rip saw

It is used for cutting the stock along the grain. The cutting edge of this saw makes a steeper angle (about 60°), whereas that of cross cut saw makes an angle of 45°with the surface of the stock.

Tenon saw

It is used for cutting tenons and in fine cabinet work. The blade of this saw is very thin and so it is stiffened with a thick back strip. Hence, this is sometimes called as backsaw. The teeth shape is similar to cross cut saw.

Cross cut saw

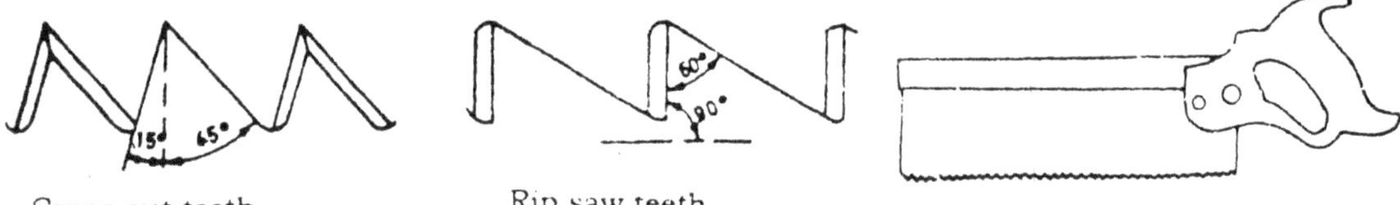

Cross cut teeth   Rip saw teeth

Cross-cut and rip saw   Tenon saw

Coping Saw

It has very small blade used for cutting small and intricate parts with curves.

Compas Saw

It has a narrow tapering blade of 250 mm long which can enter confined spaces for cutting.

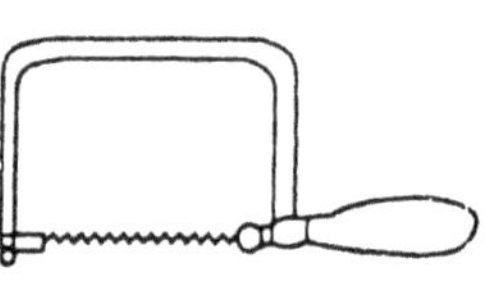

Coping Saw

Firmer chisel

Chisels are used for cutting and shaping wood accurately. Wood chisels are made in various blade widths, ranging from 3 to 50mm. They are also made in different blade lengths. Most of the wood chisels are made into tang type, having a steel shank which fits inside the handle. The firmer chisel shown in figure is a general purpose chisel and is used either by hand pressure or mallet. The blade is flat and it is specified by the width of the blade say 25mm, 20mm etc.

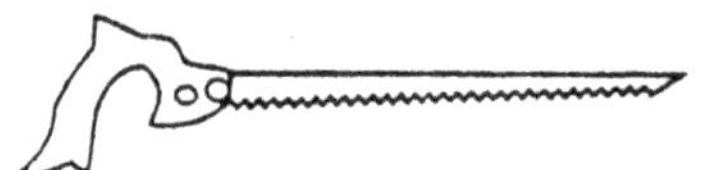

Compass Saw

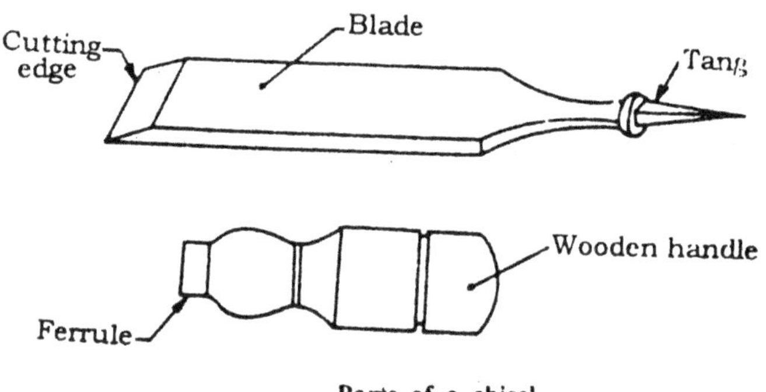

Parts of a chisel

Dove-tail Chisel

It has a blade with a bevelled back, as shown in figure, due to which it can enter sharp corners to finish them, as in dovetail joint.

Paring Chisel

Firmer or bevelled edge chisels having narrow thin blade are known as paring chisels. It is used for light work.

Mortise Chisel

These are used for cutting mortises. The cross-section of the mortise chisel is proportioned to withstand heavy blows during mortising. The cross section is also made stronger near the shank.

Firmer   Dove-tail   Paring   Mortise

Chisels

Adze

Adze is a side axe used to chap extra wood quickly by carpenters. Adze with very long handle is used to plane long palm tree barks in standing position.

Adze

DRILLING AND BORING TOOLS

Auger bit

This is used to produce long and accurate holes in wood.

Gimlet

This is a hand tool used for boring holes with hand pressure.

Twist

Scoring nib

Cutting nib

Carpenters Brace

It is used for rotating auger bits, twist drills etc, to produce holes in wood.

Auger bit

Hand drill

Carpenters brace is used to make relatively large size holes; where as hand drill is used for drilling small holes. A straight shank drill is used with this tool. It is small, light in weight and may be conveniently used than the brace. The drill bit is clamped in the chuck.

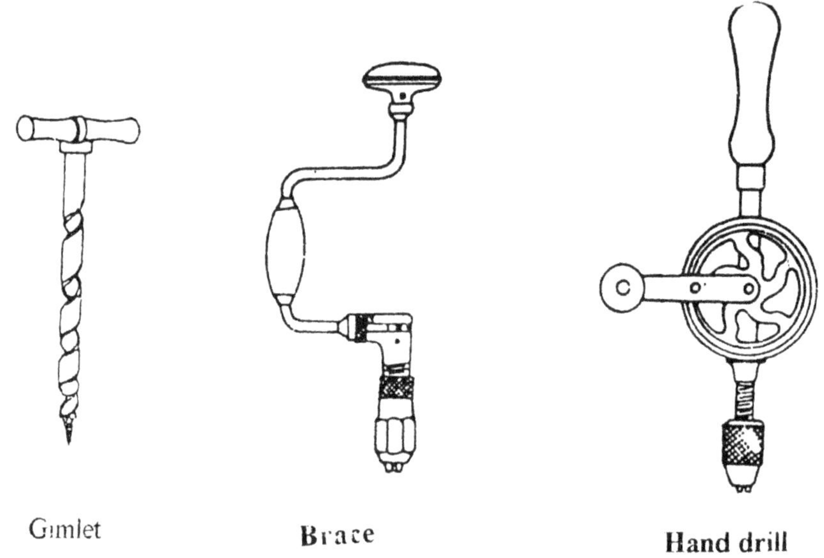

Gimlet   Brace   Hand drill

Miscellaneous Tools

Screw driver

The screw driver of a carpenter is different from the other common types, as shown in figure It has a thick blade and long handle to apply greater pressure to drive-in or remove wood screws.

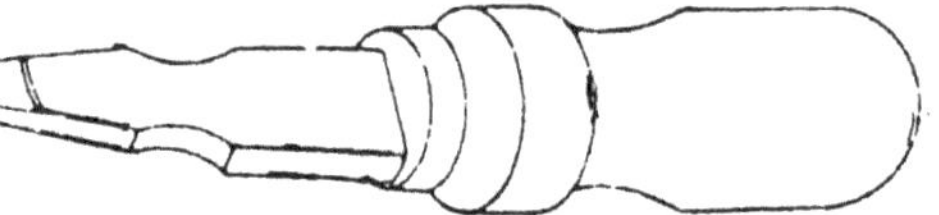

Screw driver

Mallet

It is used to drive the chisel, when considerable force is to be applied. Steel hammer should not be used for the purpose, as it may damage the chisel handle. Further, for better control, it is better to apply a series of light taps with the mallet rather than a heavy single blow.

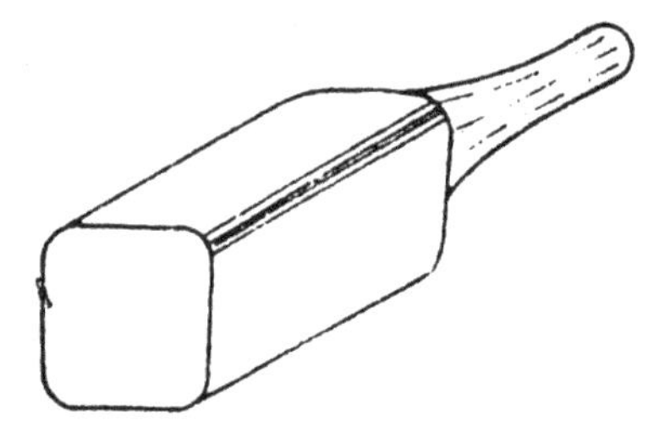

Mallet

Wood-rasp file

It is a finishing tool used to make the wood surface smooth, remove sharp edges, finishing fillets and other interior surfaces. Sharp

Wood rasp file

Tools Required : 300mm steel rule, jack plane, try-square, marking gauge, 25mm firmer chisel, marking knife or scriber, 6mm mortise chisel, hand saw, tenon saw.

Sequence of operations

1. The given reaper is checked for dimensions.
2. One side is planed with jack plane and checked for straightness.
3. The adjacent side is also planed.
4. The two surfaces are checked for sqareness with a try-square.
5. Marking guage is set and lines are marked at 30 and 45mm to mark the thickness and width of the model respectively.
6. The excess material is first chiselled with firmer chisel and then planed to correct size.
7. Now the portions for male and female pieces are marked.
8. Using tenon saw the portions to be removed are cut in both the pieces.
9. The waste pieces in 'X' are chiselled with a mortise chisel.
10. The waste pieces in 'Y' are chiselled to suit X.
11. Excess lengths on the ends are cut with a chisel.

OBJECTIVE QUESTIONS

Fill in the blanks or choose the correct word.

1. Moisture content in wood is removed by________________.
2. The three defects in wood are,____________,_________, and _________
3. __________ is used to check squareness.
4. Mortise gauge has one/two pins
5. A wood screw is specified by its __________ and __________.
6. ________saw is used for small and thin cuts.
7. __________ chisel is convenient for making dovetail joints.
8. The hand tool used for rotating auger bits is called__________
9. __________ is used for making small holes.

Exercise - 2.1

Aim : To make T-half lop joint.	**Ex. No. :**
	Date :

Sketch :

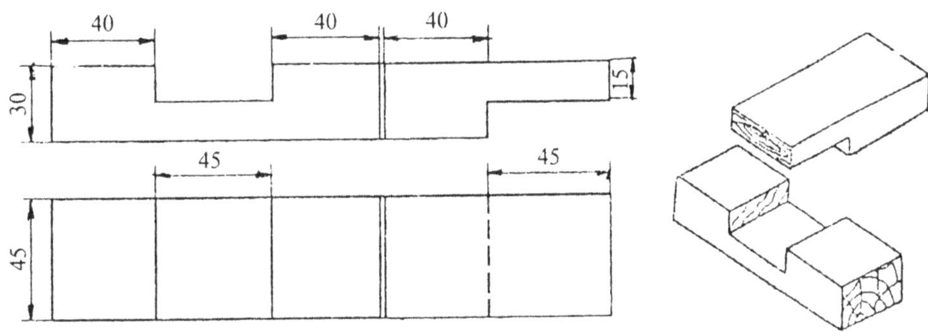

Material : Country wood

       size : 50 mm × 35 mm × 210 mm.

List of tools used :

Sequence of operations :

Safety precautions :

Staff signature

Exercise - 2.2

Aim :	Ex. No. :
	Date :

Sketch :

Material :

List of tools used :

Sequence of operations :

Safety precautions :

Staff signature

Exercise - 2.3

Aim :	Ex. No. :
	Date :

Sketch :

Material :

List of tools used :

Sequence of operations :

Safety precautions :

Staff signature

Exercise - 2.4

Aim :	Ex. No. :
	Date :

Sketch :

Material :

List of tools used :

Sequence of operations :

Safety precautions :

Staff signature

CHAPTER - 3

ELECTRICAL WIRING SHOP

INTRODUCTION

Electric wiring is defined as a system of electric conductors, components and apparatus for conveying electric power from the source to the point of use. The wiring system must be designed to provide a constant voltage to the load.

Electric power is supplied to domestic installation through a phase and a neutral forming a single phase a.c. 230v, two wire system. For industrial establishments, power is supplied through three phase four wire system, to give 440V. Figure shows the power tappings for domestic and industrial pourpose. The neutral is earthed at the distribution sub-station of the supply

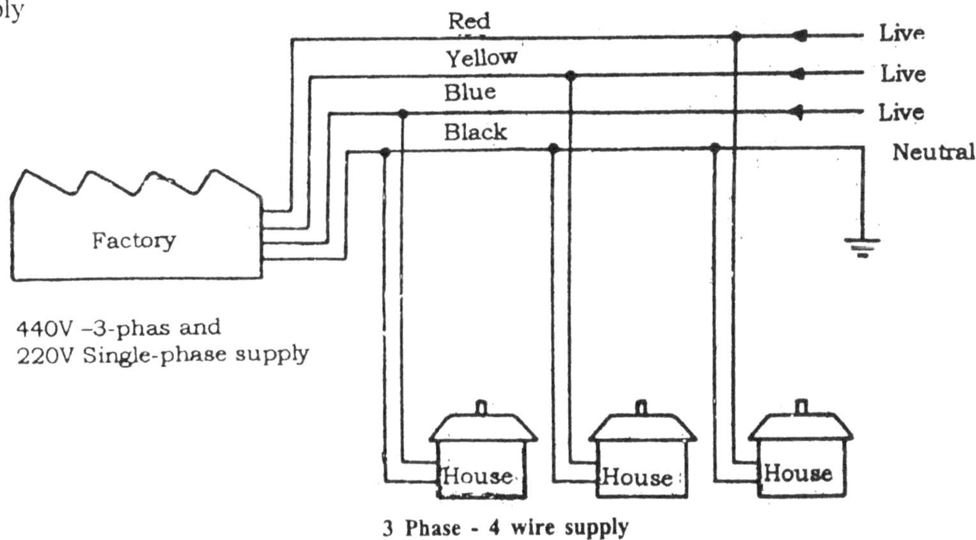

3 Phase - 4 wire supply

When supplied to domestic utilities, power is fed to a kilowatt meter and then to a distribution panel. The panel distibutes power along several circuits. It also protects these circuits from overload by safety devices like fuses or circuit breakers. The panel also serves as a main switch.

As a safe practice, all single phase devices such as switches, fuses etc, are connected to the live conductor. All electrical conductors and cables are colour coded and must be correctly connected up.

WIRES AND WIRE SIZES

An electric wire is an insulated conductor consisting of one or several strands. The insulation material is made of vulcanised India rubber (VIR) or polyvinyl chloride (PVC). The wire may consist of one or several twisted strands. A multi-core conductor consists of several cores insulated from one another and enclosed in a common sheathing.

Wire sizes are specified by the diameters of the wire, using a standard wire gauge (SWG), which also gives an idea of the current carrying capacity. The specification consists of both the number of strands and the diameter of each wire in it. For example, the specification i. Silk wire 14/36 consists of 14 strands of 36 gauge each and ii. 3/18 PVC consists of 3 strands of 18 gauge each.

WIRING METHODS

There are three types of electric circuits that are used for connecting devices or controls, to the power source, Viz, series cicuit, parallel circuit and combination of the two.

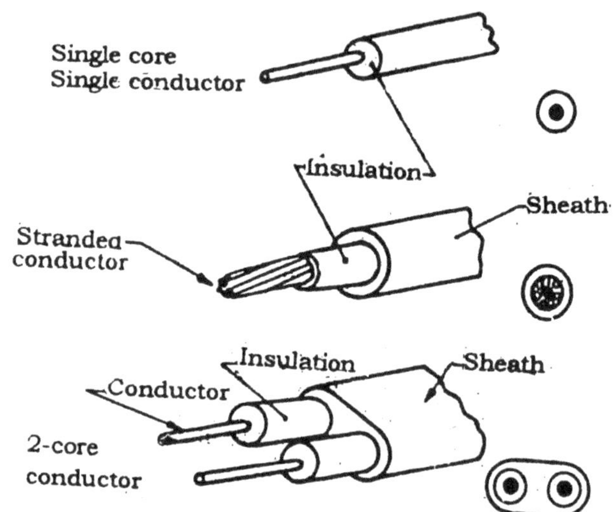

Types of Conductors

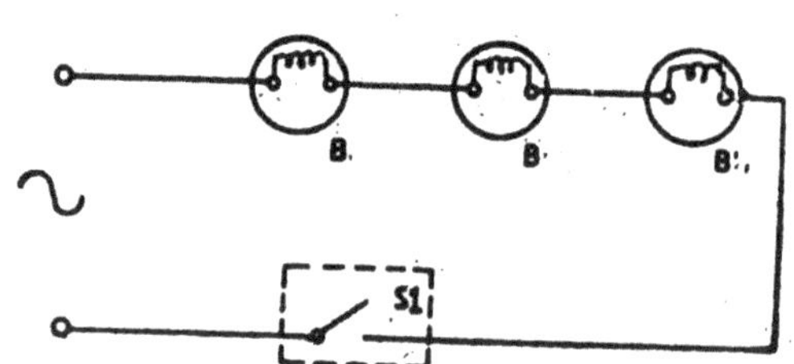

Series circuit

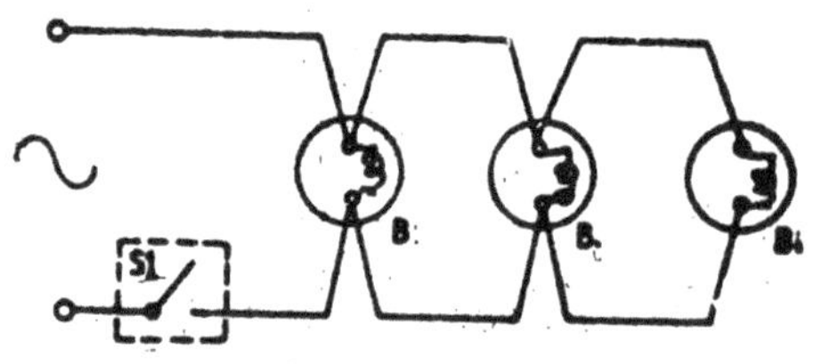

Parallel circuit.

Wiring methods

The series circuit provides a single, continuous path through which current flows. In this, the devices are connected one after another and the current flows through them until it returns to the power source. In this,even when one device breaks down, the remaining devices will not operate, because the circuit is broken. In parallel circuit, the devices are connected side by side, so that current flows in a number of parallel paths. In this type of circuit, each device is connected across the power source so that even if one device breaks down, the other devices continue to operate. Hence, this type of circuit is used in house wiring. The wires used in house wiring contain multistrand copper wires, covered with PVC insulation.

COMMON HOUSE WIRING CONNECTIONS

One lamp controlled by a one-way switch

Figure (a) shows the wiring diagram for a lamp controlled by a one-way switch. This is the normal connection one comes across in house wiring. Houever, more than one lamp may be connected either in series or parallel and controlled by a one-way switch as showr in figure (b) and (c) respectively.

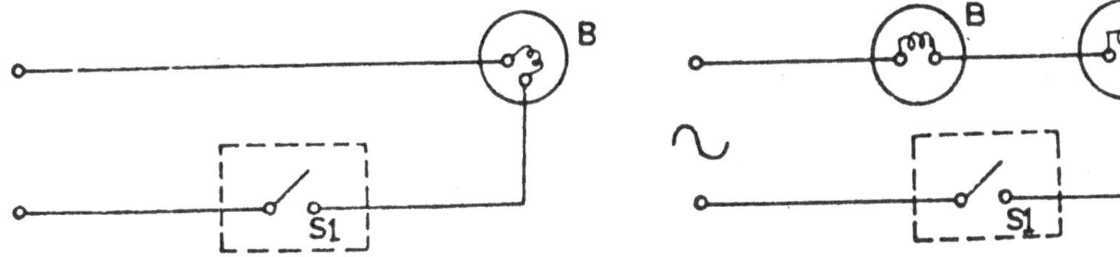

a- One lamp controlled by one way switch

b- Two lamps in series

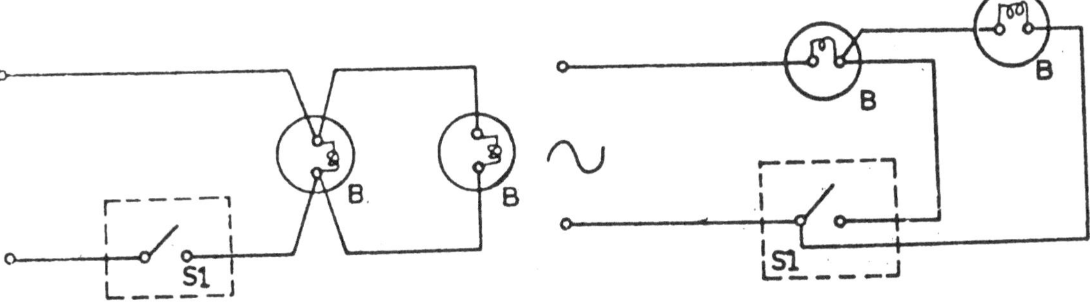

c-Two lamps in parallel

d-Two lamps in series or parallel

Common wiring diagrams

A leader finds out what more can be done. He is never out of a job.

Two lamps connected in series or parallel by a one way switch

Two lamps may be connected by a one-way switch in parallel for bright glow or in series for dull glow. This is recommended when the intensity in the room has to be controlled as shown in (d).

Two-way switching for staircase light

It is sometimes desirable to control a lamp from two different places. One may come across this situation with stair-case, bed room, long corridors or hall containing two entrances etc. This is achieved by two-way single pole switches as shown in figure.

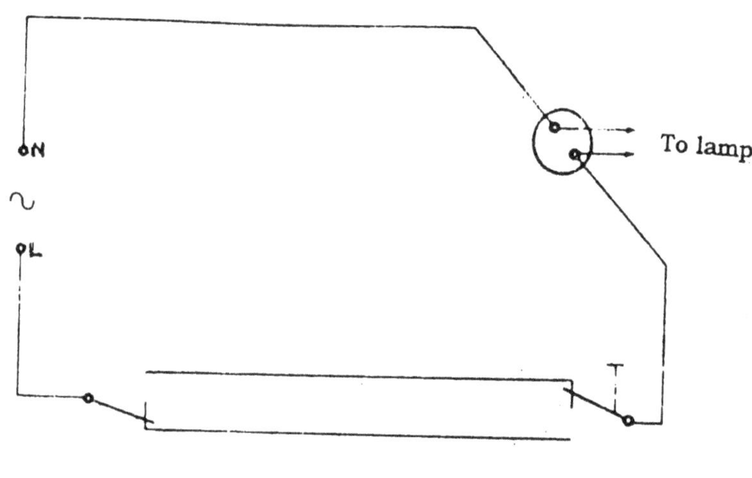

Stair case wiring

Tube Light Connections

Figure shows a typical fluorescent lamp circuit. For the lamp to operate, the switch is closed first.Current does not pass immediately through the lamp, but a potential difference (PD) occurs across it and across the starter switch. A discharge takes place which brings two contacts together. Current now flows around the circuit, heating the lamp filments. This starts the process of ionisation of gas in the lamp. When the contacts (starter switch)springs apart, a high voltage is induced in the choke. This rises the P.D across the ends of the fluorescent lamp, attracting the electons from the filaments and across it to glow.

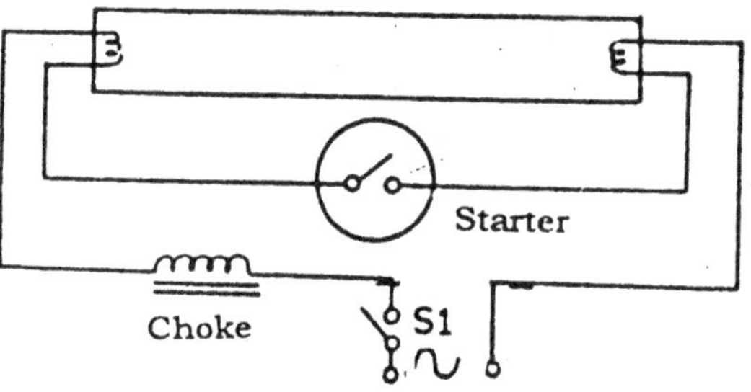

Fluorescent lamp circuit

Bell Circuit

The wiring diagram of a bell is shown below

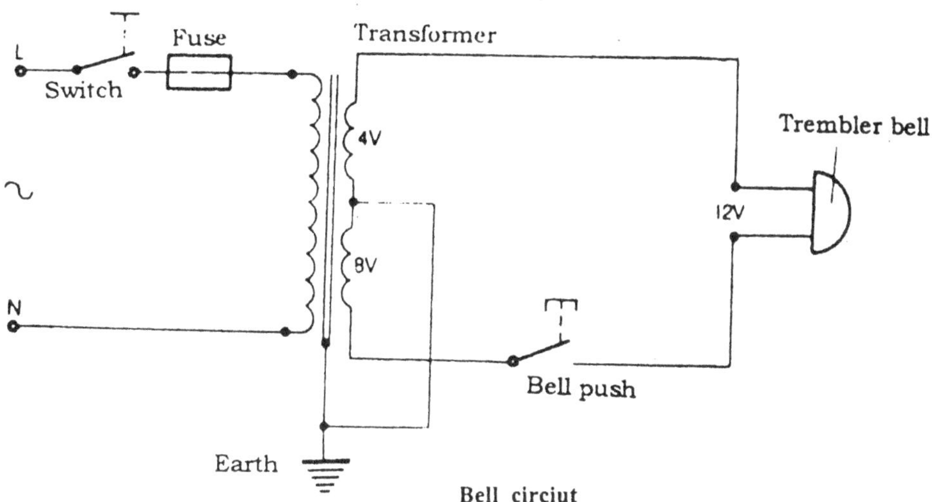

Bell circiut

Distribution Board

Figure shows typical arrangement of final circuits which may emanate from the consumer board for a domestic dwelling. Looping is done at the swithes as shown in figure.

Fuses and Circuit Breakers

These are devices designed to provide a circuit with excess current protection. In old type of distribution, panels, open link fuses, plug or cartridge fuses were used, In new panels, circuit breakers are used. When something goes wrong with an appliance or supply, the line becomes overloaded or short circuited. Then, either the fuse blows out or circuit breaker trips open, isolating that circuit or appliance. In such cases the appliance must be checked for defects or it must be ensured that there are not too many appliances in that particular circuit.

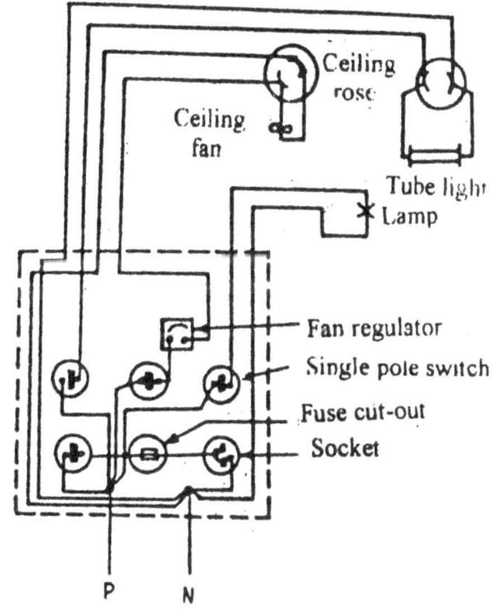

Wiring diagram of a room

Godown wiring

Figure below shows different types of fuses that are in use. Three terms are used in connection with fuses.

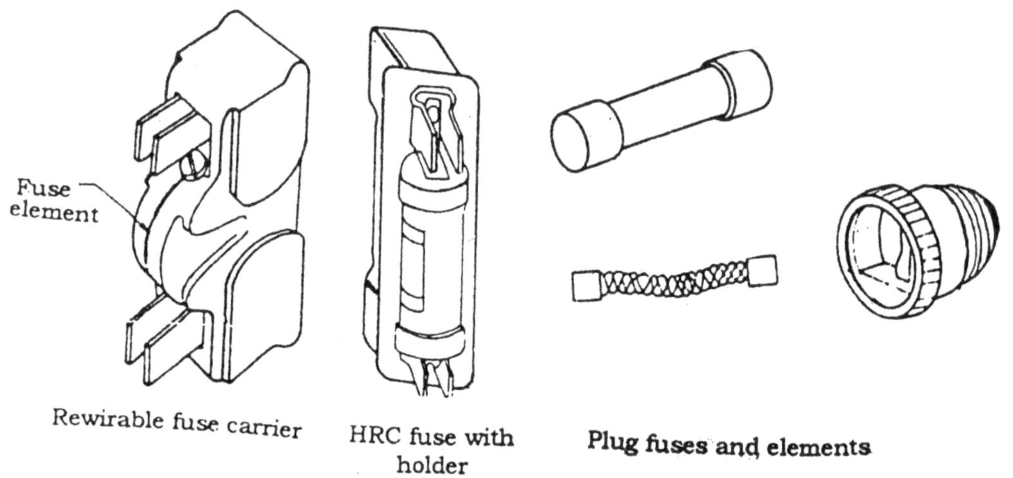

Types of fuses

i.   Current rating: This is the maximum current which the fuse will carry indefinitely without undue deterioration of the fuse element.

ii.   Fusing current: This is the minimum current which will 'biow' the fuse.

iii.   Fusing factor: This is the ratio of the minimum fusing current to the current rating, namely. fusing factor = minimum fusing current/current rating.

The rewirable fuse is a relatively cheap type of over current protective device and is still widely used despite several disadvantages. The fusing factor for rewirable fuse is about 2. Thus a fuse element rated at 10 amperes will melt (blow)at 10 x 2 = 20 amperes, which causes danger to the installation because of high fault current of twice the normal design current.

The obvious disadvantages of the rewirable type of fuses led to the development and use of cartridge and high rupture capacity (HRC) fuses. The fusing factor for this type of fuse is 1.5.

Now a days miniature circuit breakers (MCBs) are available marketed by MDS switch gears Ltd. The distribution boards are designed for locations where power demand warrants a three phase incomer but the out going circuits are only single phase as in house, shops, commercial establishments etc. The current ratings of MCBs range from 5 A to 60 A.

MDS "LOAD STOPS" are also available for protection against current shocks. Portable appliances with long trailing cables pose hazards of earth faults or electric shock - arising out of frayed or cut wires, loose connections or faulty appliances. For protection against such electric shock hazards, units with LOAD STOP RCCBs combined with metal clad plug and socket detect the earth fault and automatically cut of the supply, preventing fatal shocks.

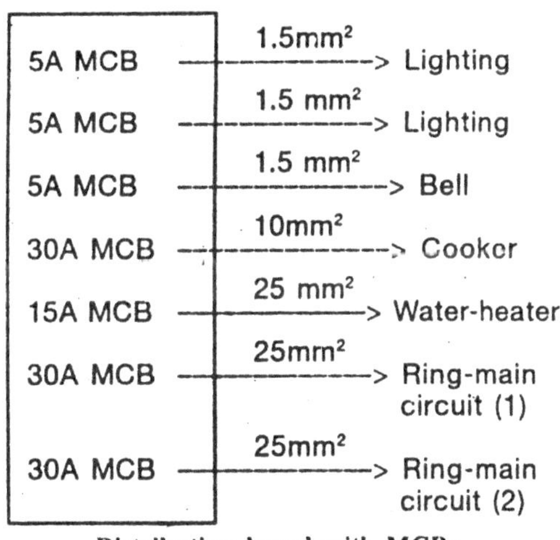

Distribution board with MCBs

Wiring diagram of a room from a distribution board is shown is above. Theoretical circuit symbols of some of the electical items is shown in Table.

Table : Symbols used in Electrical Circuits (BIS)

Aerial		Choke coil	
Earth		Cell	
Bell		Ampere meter ac/dc	
DC		Single phase	1ϕ OR $1\sim$
AC		Three phase	3ϕ OR $3\sim$
Transformer	H.T ‖ L.T	Phase sequence	RYB
Full-wave rectifier		Cartridge fuse	
Star connection		Fan regulator	OFF ON
Delta connection		**Three-plate ceiling rose**	
Volt meter ac/dc		Batten lamp holder	
Fan		One-way switch	
Exhaust fan		Two-way switch	

PARTS OF A CABLE

1. Conductor,
2. Insulation covering, and
3. Protective covering.

STRANDED CABLE

Electrical energy is supplied from the generating station to the consumer by means of cables. The conductor of the cable is of two types.

1. Solid conductor, and
2. Stranded conductor

In a solid conductor cable, there is only one conductor. But in stranded conductor cable, it is made of a number of strands of wires of circular cross-section so that it is flexible.

The number of strands used in a cable are 3, 7, 19, 37, 61, 91, 127 or 169. These numbers are chosen to give a circular shape to the conductor of the cable. In a three strand cable, two strands are twisted around the third strand. Cables are stranded to increase the current carrying capacity of the cable as compared to solid conductor because it has larger heat-radiating and conducting surfaces and therefore allows a higher current for the same temperature.

Advantages of stranding cables

1. Thin wires can be used which increases the flexibility of the cable.
2. Provides ease in handling during installation and erection work.
3. Facilitates making joints.
4. If a conductor breaks, there are other conductors to which the current can pass and thus there is no complete break down.
5. It provides ease in soldering joints.

Core of a cable

The core of a cable is single conductor of a cable with its insulation but not include any mechanical protective covering.

Colour code for cores

For identification of multi-core cables having wires of different polarities, the cores are allotted different core colours.

In a three-phase, four-wire system, the three phases 1, 2 and 3 are given the core colours **Red, Yellow and Blue** respectively. For a neutral wire, the colour of the insulation is **Black,** and **Green** for earth wire. These core colours should always be kept in mind while connecting conductors in multi-core cables.

Size of a cable

The size of a cable depends on the size of the conductor. It is determined by one of the following :

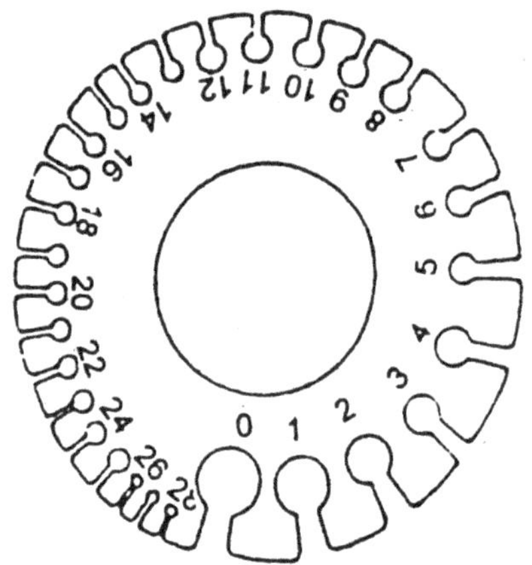

Standard wire gauge

1. with a standard wire gauge,
2. according to the diameter of the conductor, and
3. according to the cross-sectional area of the conductor.

Figure shows a standard wire gauge. The gauge of a wire is determined by trial and error by inserting in the slots. It is specified by the number of the SWG indicated. If a stranded cable is of 3/20, it means that the conductor is made of copper and there are three strands in a cable each of 20SWG. Similarly if a conductor of aluminum cable has a cross-sectional area of 25mm², it can have a 7/2.24 stranded conductor. It has seven strands in a cable and each strand has a cross sectional area of 4mm².

Types of cables

The following cables are used for different types of wiring installation.

1. V.I.R. (Vulcanized Insulation Rubber) cable.
2. C.T.S (Cab Type Sheathed) cable
3. P.V.C. (Poly Vinyl Chloride) cable.

The sizes available in copper conductors are

1/18, 3/22, 3/20, 7/22, 7/20, 7/18….. etc. Similarly, in aluminium the conductor sizes are given in mm as 1/1.4(1.5mm^2), 1/1.8 (2.5mm^2), etc.

Of all the cables the PVC is extensively used because the polyvinyl chloride insulation serves both as insulation covering as well as mechanical protective covering. It resists the action of acids, alkalis and variation in atmospheric temperature.

Its use is restricted where there is possibility of the temperature being very high as it softens and at low temperatures where it becomes brittle. This cable is replacing the CTS cable. It is used for indoor wiring and panel wiring and is available in the same sizes as the VIR cable. No tinning of either copper or aluminium conductors is necessary to harden them by adding sulphor because the PVC insulation itself is very hard.

Voltage grade of cable

The working voltage of wiring decides the grade of the cable and the thickness of the rubber insulation covering of the cable. Therefore, cables are manufactured in two different grades namely 230/400V and 650/1100V grade cables are used for industrial power wiring.

Selection of cable for wiring

The following points should be considered while selecting a cable for a wiring installation :

i Effect of atmosphere : The insulation of the cable should not be effected by the surrounding atmosphere condition. For example, a weather-proof cable is used for open places, a Tropodure cable is used for oil mills and a Lead-covered cable is used in chemical plants as the acid fumes can destroy the insulation of cables.

ii. Maximum voltage of the circuit : The grading of the cable should be equal to the maximum working voltage of the circuit.

iii. Full load current of the circuit : The current rating of the cable must be at least to carry the full load current of the circuit. Table gives some of the details.

Table : Current rating capacity of VIR and PVC insulated cables.

Sl No.	Stranded Cu. conductor	Stranded Al. conductor	Area mm^2	Current rating (amps)
1.	1/0.044(1/18)	-	-	5
2.	3/0.029(3/22)	1/1.40	1.5	10
3.	3/0.036(3/20)	1/1.80	2.5	15
4.	7/0.029(7/22)	-	-	20
5.	-	1/2.24	4	20
6.	7/0.036(7/20)	-	-	28
7.	-	7/1.70	16	43
8.	19/0.044(19/18)	-	-	62
9.	-	19/1.8 or 7/3.00	50	91

Systems of Wiring

The following are the various systems of domestic and industrial wiring.

1. Cleat wiring system.
2. Wooden casing and capping wiring system.
3. Lead sheathed wiring system with clips.
4. CTS wiring system with clips.
5. Conduit pipe wiring.
6. PVC casing and capping wiring system.
7. PVC pipe wiring system.

Now a days only PVC pipe wiring is done extensively for domestic use. Cables are connected direct for heavy loads in industry. PVC casing and capping wiring system has replaced wooden casing and capping where external wiring is done.

Other wiring accessories

1. Single pole one-way flush mounting switches of 5 or 15A, 250V.
2. S.P two-way flush mounting switch 5A, 250V. These are also known as piano switches.
3. Two or three pin flush mounting wall socket 5A, 15A.
4. Plug tops of two pin and three pin 5A & 15A.
5. Batten lamp holders.
6. Pendent holders.
7. Ceiling roses-2 plate and 3 plate types.

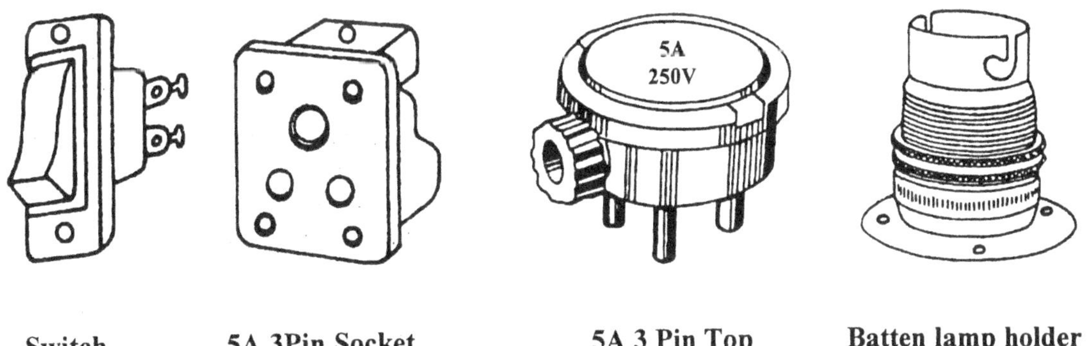

| Switch | 5A 3Pin Socket | 5A 3 Pin Top | Batten lamp holder |

House hold electric appliances

1. Automatic Electric Iron

The exploded view of an automatic electric iron is shown with a thermostat which is connected in series with the heating element.

Thermostat is a switch which operates automatically due to the variation of heat produced around it. It functions on the principle that different metals have different rates of expansion when heated. The coefficient of expansion of brass and iron are different. If a strip is made of bimetal say brass and iron and heated beyond a certain temperature it will bend down wards.

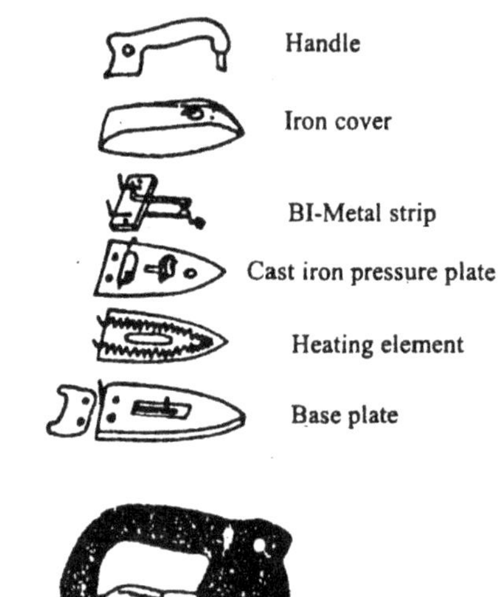

Automatic electric iron

The above principle is used in operating the thermostat. It has two contact pints, namely, the fixed contact point and movable contact point. The movable contact point of the thermostat consists of a bimetal strip which is normally in contact with a bimetal strip which is normally in contact with another strip of metal having fixed contact. The thermostat is connected is series with the heating element as shown in figure.

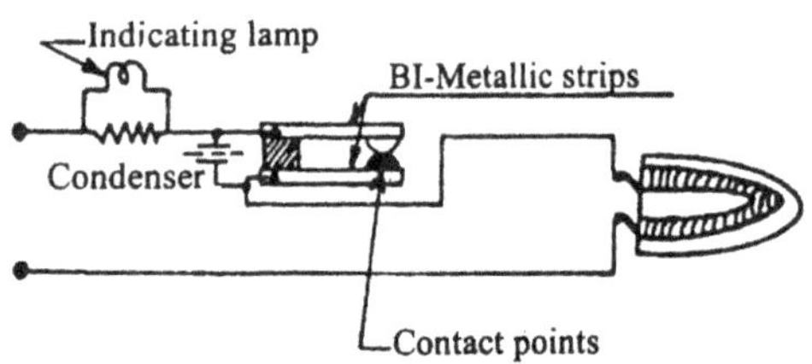

Thermostatic control of an electric iron

A condenser is connected across the contact point of the bimetal strip. Its function is to minimize the sparking at the contact point. An indicator lamp of 2.8V is also fitted which is connected in series with the element. The automatic irons are of 750 watts.

2. Storage water heater :

It consists of a double walled reservoir known as the cistern, a heating element and a thermostat as shown in figure.

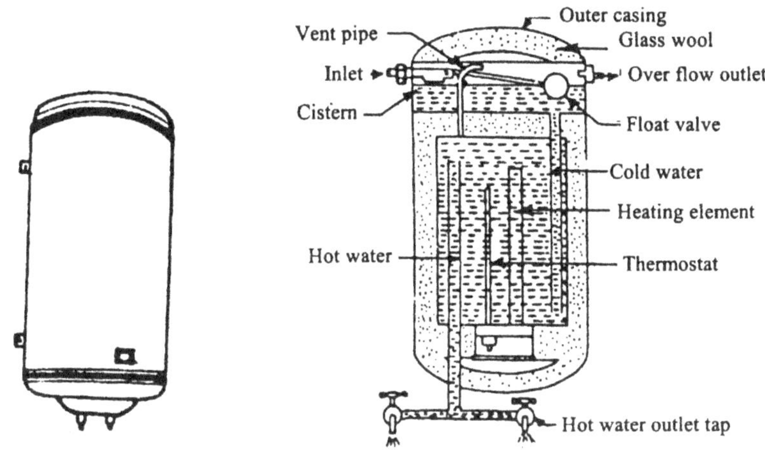

Pressure type water heater

SAFETY PRACTICES

1. When closing the electric switch, always grasp the switch, by the insulated handle.

2. Do not run too many electrical items from one point.

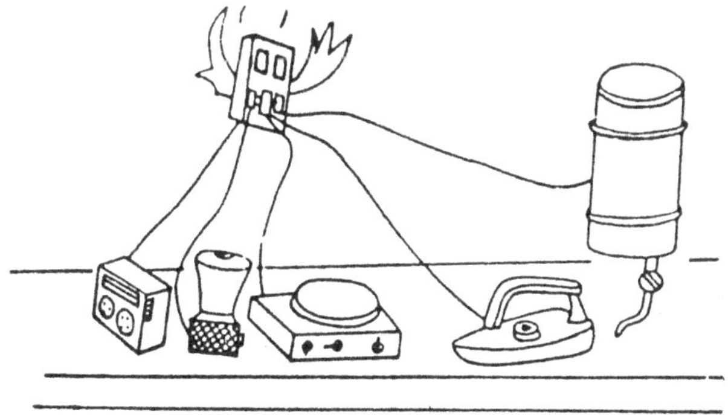

Over load of current drawn at
single point results in burning of
the cables and fire accident.

Think and thak God for what we have.

Electrical tools improperly
grounded may result in electric
shock.

3.   Use fuses and circuit breakers of proper capacity, so as to interrupt the current before it becomes dangerous.

4.   Disconnect the units to be repaired free from power supply and make sure that they might not be energized while the repair work continues.

5.   Do not pour water to put off fires in electric wires and electric equipment. You will be subjected to electric shock or you will be electrocuted. Use sand to put off fires in electric items.

6.   Whenever there is power failure, put-off the power supply to all equipment, in order to prevent spontaneous recovery.

7.   Never remove a plug from an outlet by pulling the cord. Always pull by the plug.

8.   While testing, always keep one hand in the pocket. If both the hands are in contact with a circuit, a current will flow across your body and is more dangerous.

9.   Electricity has no respect for ignorance. Do not apply voltage or turn on any device until it has been properly tested.

10.   Check the earth connection before switching on portable equipment like drill guns etc.

Exercise - 3.1

Aim : To connect two lamps in series.	**Ex. No. :**
	Date :

Connection diagram :

Two lamps in series

Material :

List of tools used :

Sequence of operations :

Safety precautions :

Staff signature

Exercise - 3.2

Aim :	Ex. No. :
	Date :

Sketch :

Material :

List of tools used :

Sequence of operations :

Safety precautions :

Staff signature

Exercise - 3.3

Aim :	Ex. No. :
	Date :

Sketch :

Material :

List of tools used :

Sequence of operations :

Safety precautions :

Staff signature

Exercise - 3.4

Aim :	Ex. No. :
	Date :

Sketch :

Material :

List of tools used :

Sequence of operations :

Safety precautions :

Staff signature

CHAPTER - 4

SHEET METAL SHOP

INTRODUCTION

Tin smithy deals with making of metal boxes, cans, funnels and ducts from flat sheet metal. In this, the development is drawn on the sheet metal and cut and folded, to form the required shape of the object. For successful working in the trade, one should have a thorough knowledge of projective geometry, particularly of the development of surfaces.

The common hand tools used in sheet metal work are, steel rule, usually of 60cm length, wire gauge, dot punch, trammels, scriber, ball peen hammer, straight-peen hammer, cross-peen hammer, mallets, snips and soldering iron.

Trammels

Sheet metal layout requires marking of arcs and circles. This may be done by using the trammels, shown in figure. The length of the beam decides the maximum size of the arc that can be scribed.

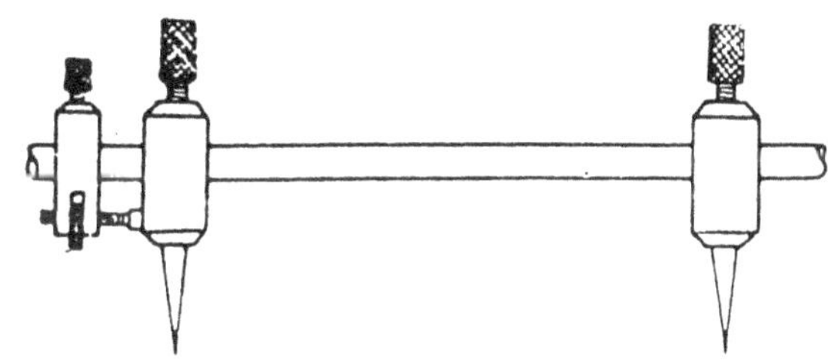

Trammel

Wire Gauge

The thickness of sheet is referred in numbers known as standard wire gauge (SWG). The gaps in the circumference of the gauge are used to check tha gauge number as shown in figure. Some of the standard wire gauge numbers with corresponding thicknesses are as follows.

Table 1

SWG No	Thickness mm
10	3.20
12	2.60
14	2.30
16	1.60
20	1.00
22	0.70
24	0.65
26	0.45
30	0.30

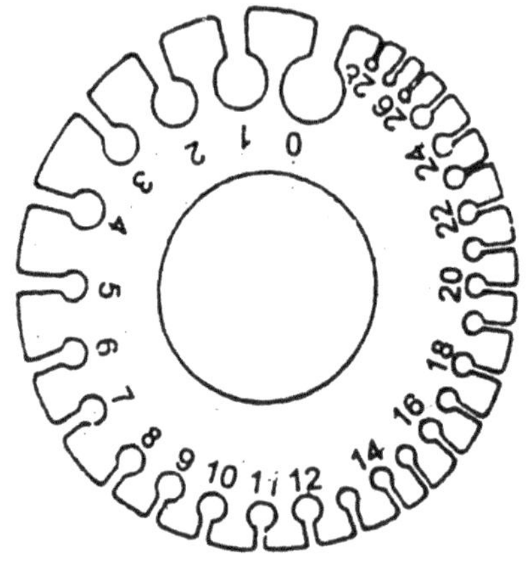

Standard wire gauge

Bench Shears

Sheet metal may be cut by shearing action. Figure shows a bench shear. In this, the force is applied through a compound lever, making it possible to cut sheet metal upto 4 mm thick. The chopping hole can shear a mild steel rod upto 100 mm diameter.

Snips

Snips are hand shears, varying in length from 200mm to 600mm, 200mm and 250mm being the lengths commonly used. Figure shows straight and curved snips. The straight snips are used for cutting along outside curves and straight lines and curved snips or bent snips are for trimming along inside curves.

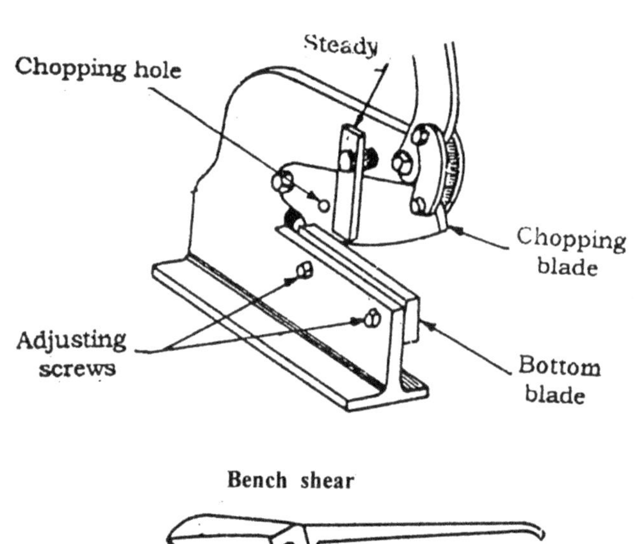

Bench shear

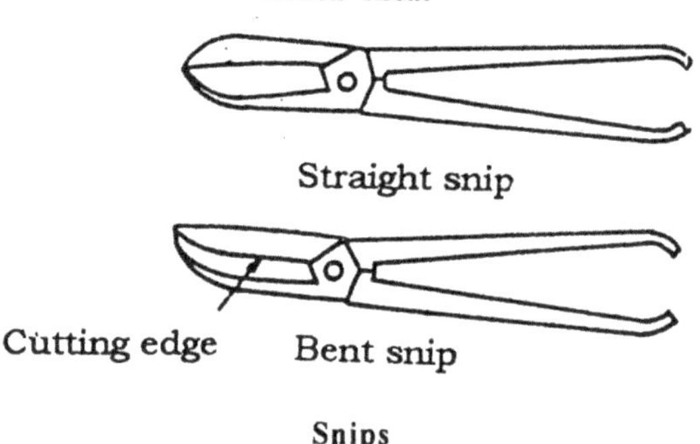

Snips

Quality is never an accident - it is the outcome of brilliant effort.

Hammers

Light weight hammers and mallets are used in sheet metal work.

Ball peen hammer has a cylindrical, slightly curved face and a ball head. It is a general purpose hammer, used mostly for rivetting in sheet metal work. Cross peen hammer shown, has a tapered peen end and is perpendicular to the handle. Because of this, it can reach aukward corners as shown.

Straight peen hammer has the peen end similar to the cross peen, but it is positioned parallel to the handle which can be used conveniently for certain operations of folding. Mallet is ued for bending and folding work. It is light in weight, covers more area and does not dent the work.

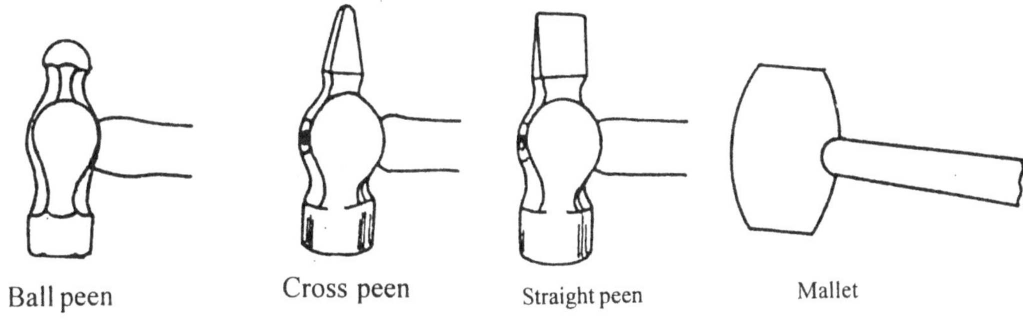

Hammers

Stakes

Stakes are nothing but anvils, which are used as supporting tools and to form, seam, bend or revet sheet metal objects. These are available in different shapes and sizes as shown in figure to suit the requirements of the work. They are made from wrought iron, faced with steel.

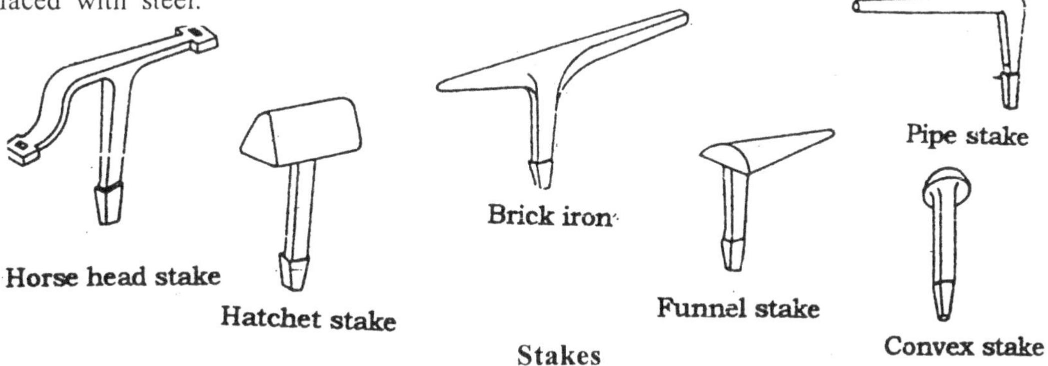

SHEET METAL JOINTS

Various types of joints are used in sheet metal work, to suit the varying requirements. Some commonly used sheet metal joints and folded edges are shown in figure. These are self secured joints, formed by joining together two pieces of sheet metal and using the metal itself to form the joint. These joints are to be used on sheets of less than 1.6mm thickness.

Sheet Metal Screws

Sheet metal screws are used in sheet metal work to join and nstal duct work for ventilation, air conditioning etc. These screws are also known as self-tapping screws since they cut their own threads. Sharp and blunt pointed screws are used generally, to join light material. The hole size should be equal to the root diameter of the screw.

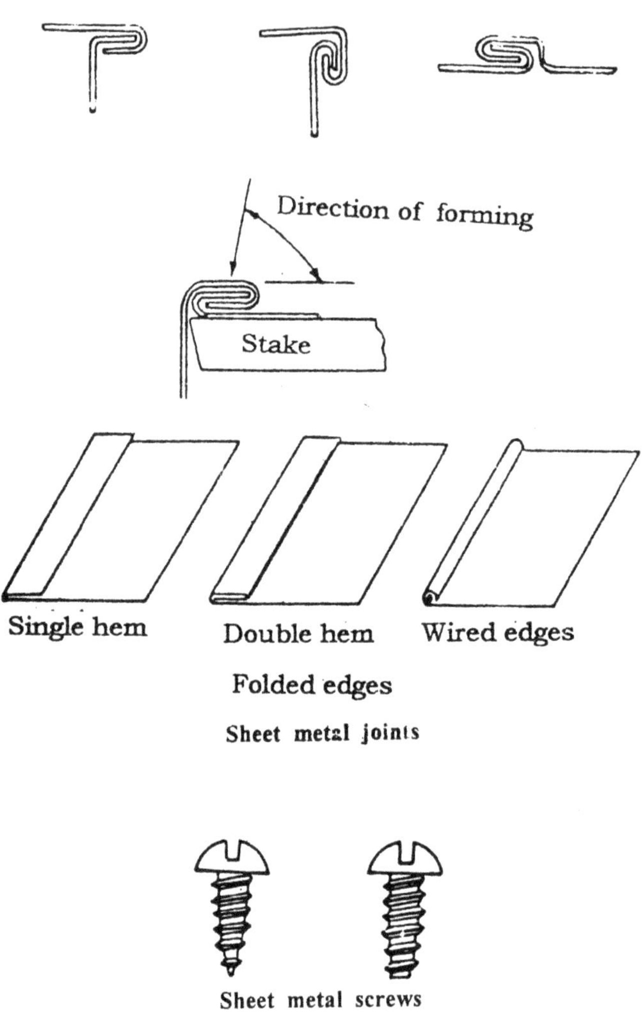

Rivetting

Rivets are used to fasten two or more sheets of metal together. It is the common practice to use rivets of the same material as that of the sheets being fastened. Tinmen's rivets with flat heads are used on sheet metal work. For successful rivetting operation, the selection of proper size and spacing of rivets is essential.

Light sheet metal is usually punched for rivetting, while the heavier sheets are drilled. A solid punch,rivet-set,snap and hammer are the tools required for the job.

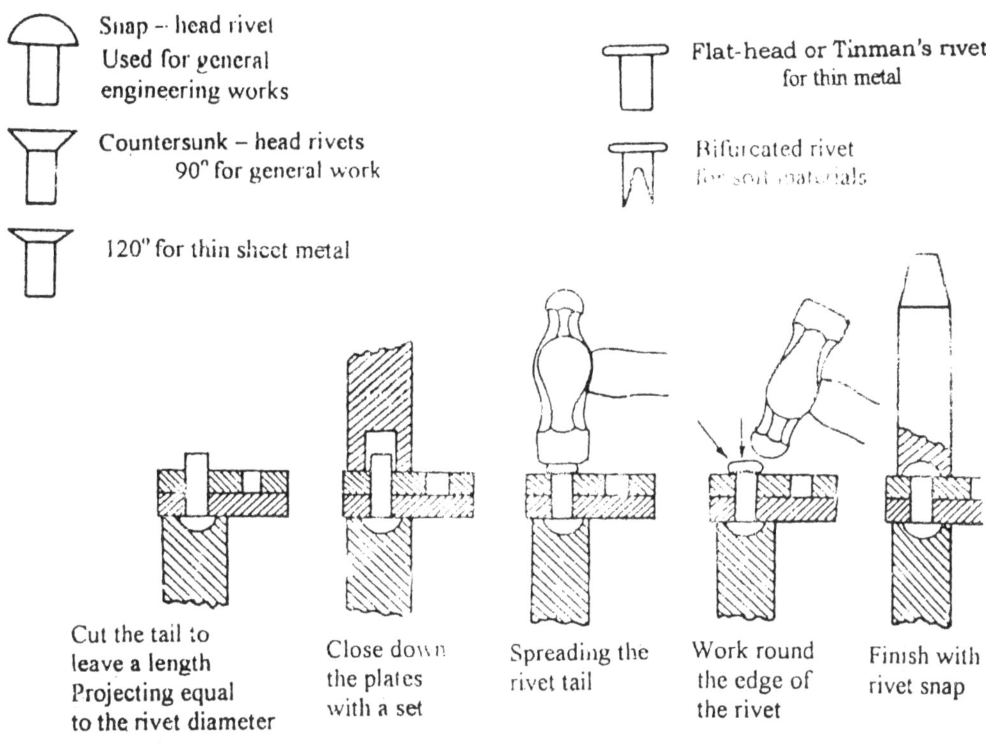

Riveting

The following rules must be observed while rivetting sheet metal work:

1. Diameter of rivet = 1.2 x thickness of one plate

2. Distance from the edge of the hole to the edge of the plate = 1.5 x diameter of rivet

3. Length of the projecting shank = diameter of rivet

For better results, it is advisable to drill all the holes in one plate and only one in the other, secure the plates with one rivet and then complete the dilling of the remaining holes.

Figure shows the stages of rivetting in sheet metal work.

Sheet Metal Layout

In sheet metal work, it is required to lay the full size pattern on the metal sheet, so that when it is cut along the pattern and folded or bent, it will result in the required object. This laying-out of the complete surface on the metal sheet is known as the development of the surface of the object. However, while laying-out, it is necessary to provide excess material for the joints known as allowance. Figure shows the layouts of certain objects and the shapes obtained when formed.

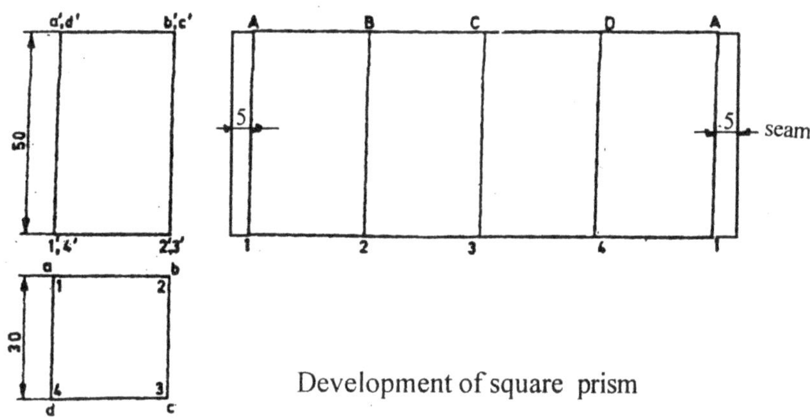

Development of square prism

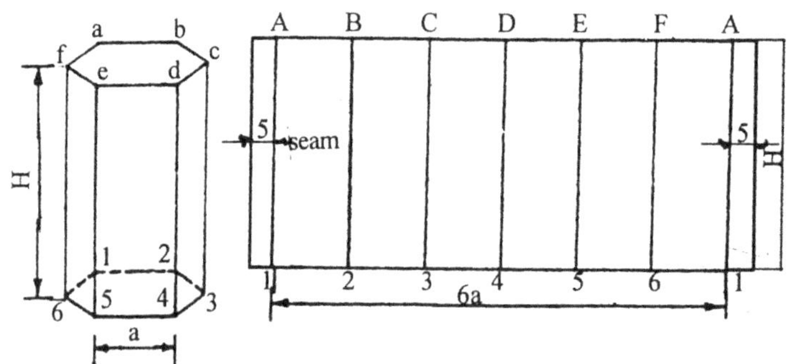

Development of hexagonal prism

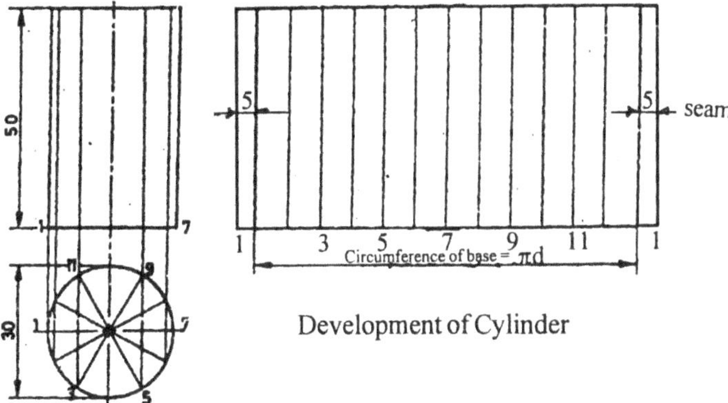

Development of Cylinder

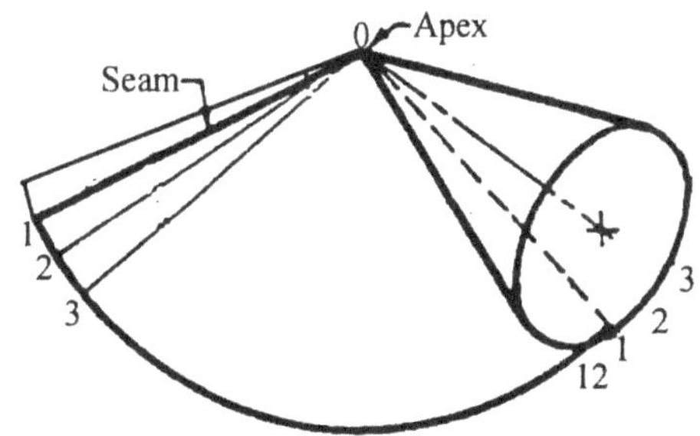

Layout of a Cone

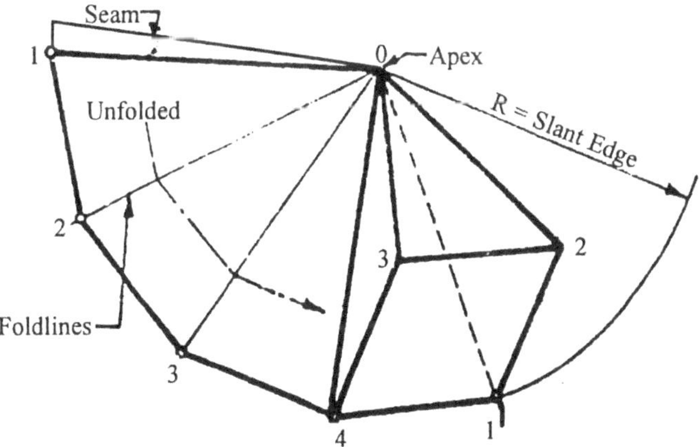

Layout of a Square Pyramid

Plan your work and work your plan.

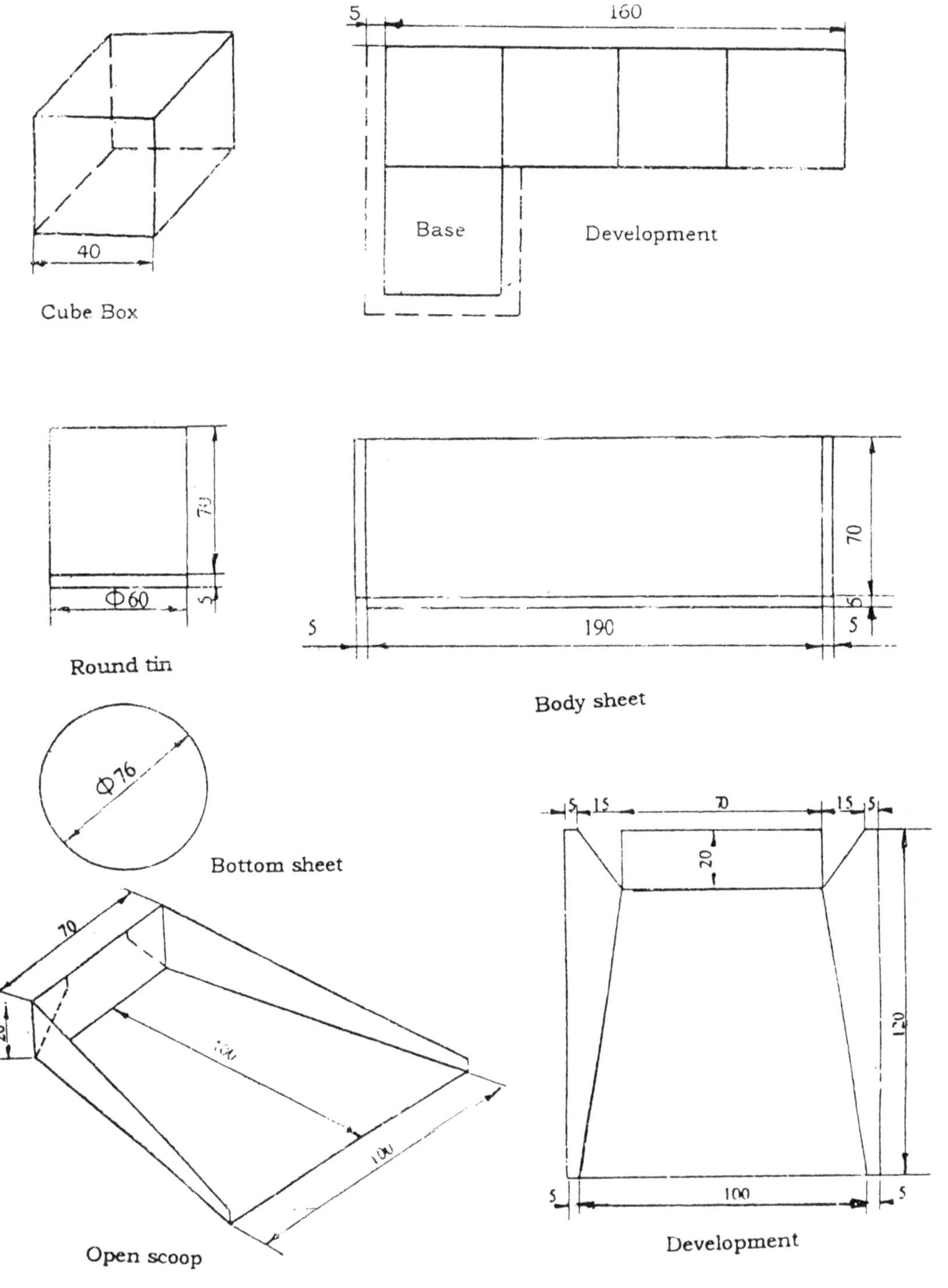

40
Cube Box
5
160
Base
Development
70
Φ60
5
Round tin
5
190
5
70
5
Body sheet
Φ76
Bottom sheet
70
20
100
100
Open scoop
5 15
70
15 5
20
120
5
100
5
Development

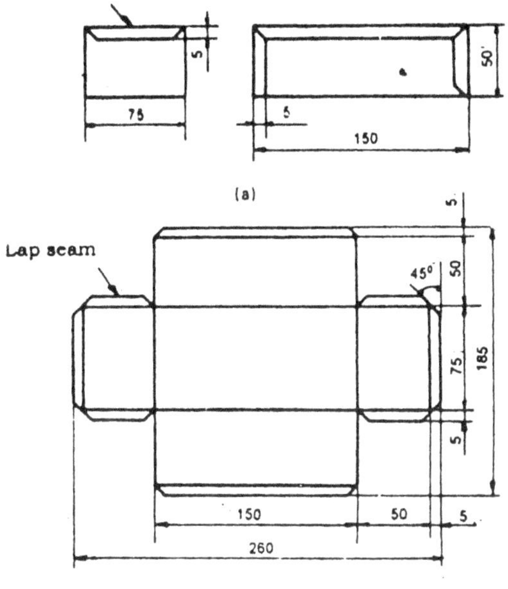

(a)

Rectangular tray

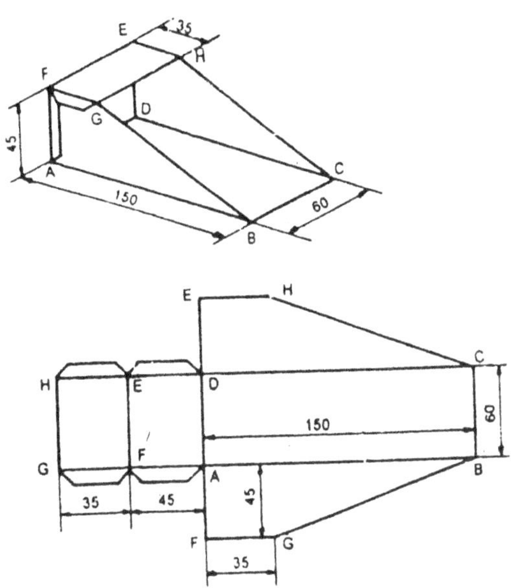

Rectangular scoop

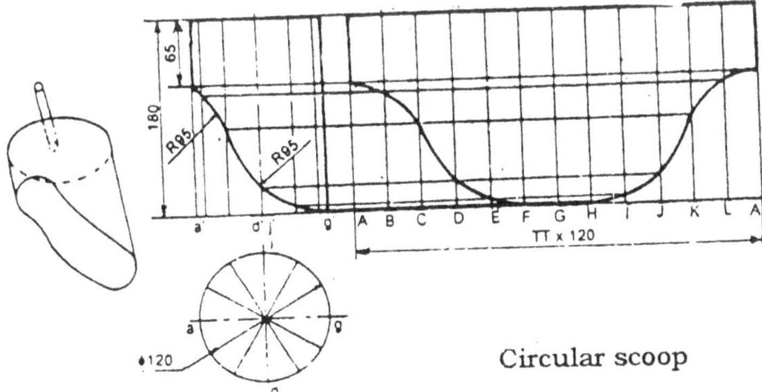

Circular scoop

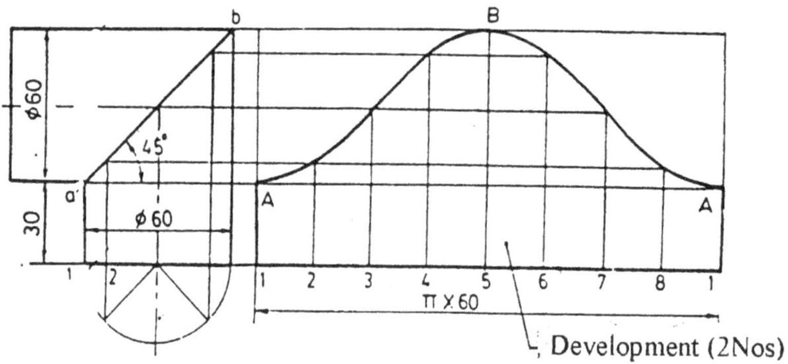

Development of a 90° Elbow

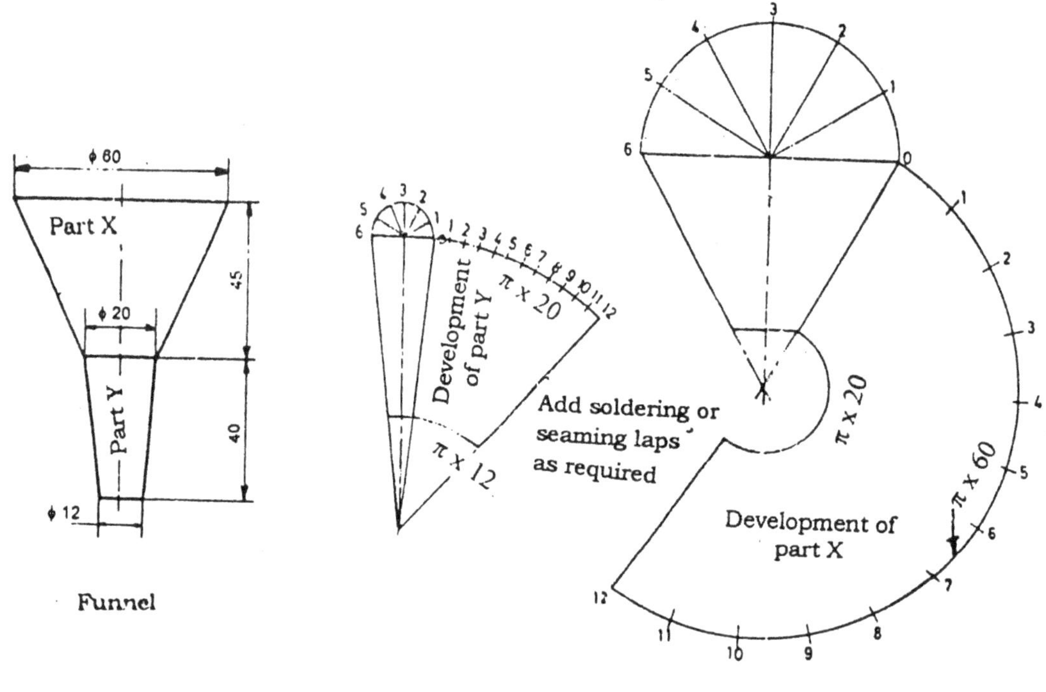

SOLDERING

Soldering is one method of joining two pieces of metal with an alloy that melts at a lower temperature than the metals to be joined. For a good job, the metals to be joined must be free from dirt, grease and oxide. Solder is made of tin and lead, usually in equal proportions. It comes either in the form of wire or bar.

Solder with 60% Tin and 40% Lead is known as tinman's solder with a melting temperature of 183°C. Plumbers solder has 40% Tin & 60% Lead and melting temperature of 234°C. This is used for joining lead pipes or lead sheathing of electric cables.

Silver solder has a composition of 34% Cu, 50% Ag and 16% Zn. With a melting temperature of 600°C. Silver soldering is done with acetylene flame with small nozzle.

Flux is used to remove oxides from the metal and to make the solder flow smoothly. An acid flame is used for black iron and galvanized iron. A **rosin flux** is used for tin plate, copper and electric wire.

Soldering Iron

Soldering requires a source of heat. A common method of transmitting heat to the metal surfaces is by using a soldering iron. Figure shows the soldering iron used for the purpose. The working end of this tool is made of copper, which is a good conductor of heat.

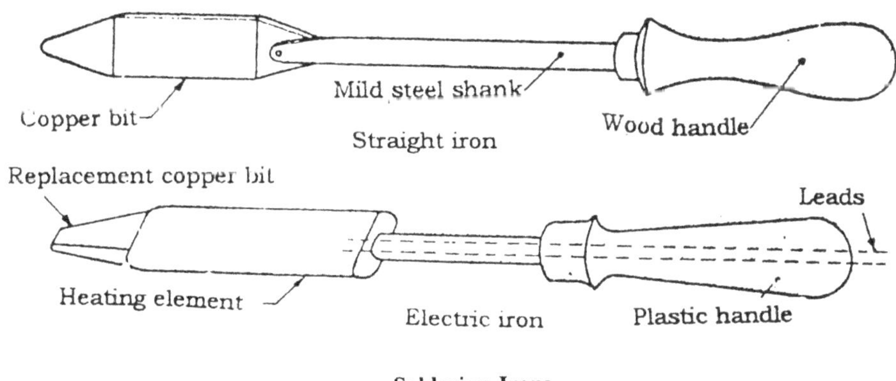

Soldering Irons

To do a good job, the soldering copper must be tinned. Also, when it is sufficiently used or over-heated, the tinned point will be covered with oxide formation, preventing the flow of heat to the solder. Hence, the point must be tinned again.

Method of soldering

The following are the stages involved in soldering work:

1. Clean the surfaces to be soldered.

2. Keep the surfaces to be joined, close together.

3. Apply a thin layer of flux with a brush.

4. Heat the soldering copper to proper temperature.

5. Tack the seam by applying solder at several points.

6. Begin at one end and move the copper bit slowly, adding solder as needed.

Note: (i) Do not handle the soldered job, until the solder is partially cooled.

(ii) Do not melt the solder on the copper bit itself.. Instead, heat the metals to be joined to melt the solder.

BRAZING

For joining non-ferrous metals such as silver, copper and brass, silver solder is used, while ferrous materials are invariably joined by a brass spelter. This process is known as brazing, also called as hard soldering. For brazing to be done, it is essential that the metals to be joined have a higher melting point than the solder or spelter used. The principle of brazing is very much similar to that of soldering. However, in this, a welding torch or blow lamp is used to heat the metal. Brazing solder with a composition of 60% Cu and 40 Zn has a melting temperature of 850°C. Borax is used as a flux for brazing. The following are the stages involved in brazing:

1. Make a tight fitting joint. Where necessary, file a groove along the joint for the solder to run into.

2. Apply flux with a brush.

3. Preheat the area until the flux dries out. Then heat the joint until it is bright red.

4. Apply the solder when it begins to melt; move the torch along the joint for the solder to flow evenly.

5. Clean, file and smoothen the joint.

Note: In soldering, also known as soft soldering, the joint is strong only when certain means such as rivetting or folded or grooved seams are used. In hard soldering, the strength of the joint depends on the area of contact between the surfaces to be joined.

SAFE PRACTICES

1. Use hand leather gloves while handing heavy sheets.

2. Avoid pulling or feeling the cut portion by hand while cutting with snip.

3. Follow the usual precautions necessary for handling hot objects, as soldering is a process requiring heat.

4. Ground properly, the exposed metal parts of electrically heated soldering irons.

5. Place the electric iron on fire proof holder and never allow it to be on a table when it can come in contact with the combustible material.

6. After performing the work, wash hands thoroughly before eating, as solders and fluxes contain material which should not enter the stomach.

7. Be cautious and avoid inhalation of cadmium fumes which are toxic and are present in greater quantity when plated materials are brazed.

8. Avoid breathing fumes from the solmmoniac as they cause head aches and injure the lungs.

To make a square tin as shown in figure.

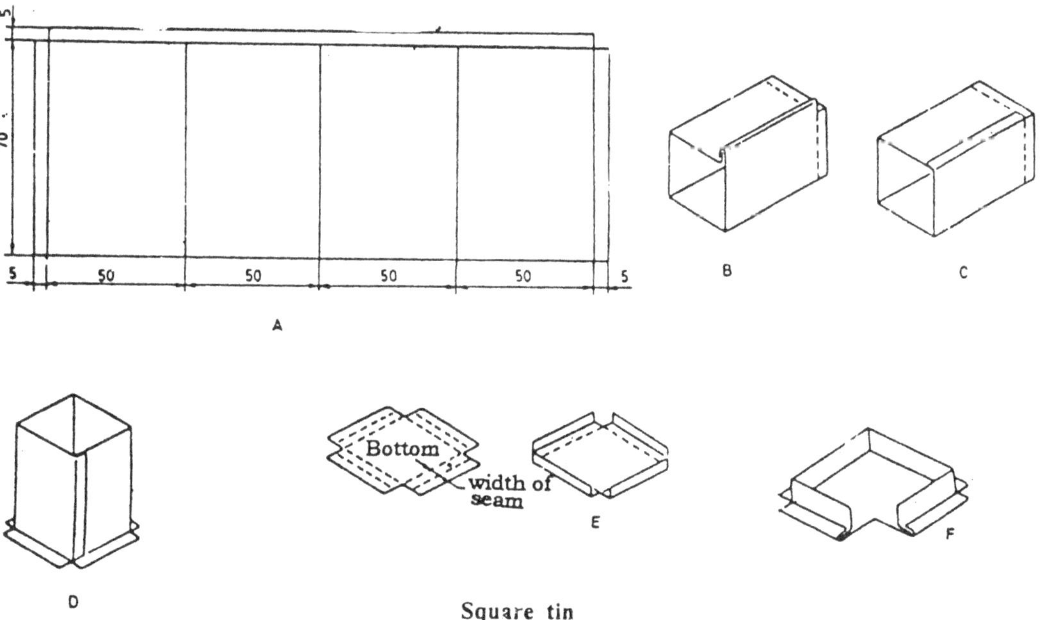

Square tin

Tools used : Divider, try-square, scriber, straight snip, 200mm steel rule and mallet.

Sequence of Steps

1. The layouts of the body and the bottom are drawn on the tin sheet and cut as shown in Fig. (a).

2. The body is folded as shown in Figs. (b) and (c).

3. The flanges are bent to receive the bottom as shown in Fig. (d).

4. The bottom is bent as shown in Fig. (e).

5. The bottom and body are placed in position and folded as shown in Fig. (f).

6. It is again folded-up to complete the joint.

7. Soldering is done if leak proof joint is required.

OBJECTIVE QUESTIONS

Fill in the blanks.

1. Sheet-metal layouts are also known as ____________________

2. Galvanised iron sheet is a steel sheet coated with __________

3. Beading is added to sheet-metal objects for adding__________

4. Sheet metal can be cut by hand with ______________

5. To measure the thickness of sheet, __________ is used.

6. __________ is used to draw an arc of very big radius.

7. The hammer used to shape sheet metal is known as __________

8. __________ are used to support the sheet while shaping.

9. 18 and 8 stainless steel sheet contains 18 per cent ______________ and 8 per cent ______________ .

10. As the gauge number increases, the sheet thickness ____________.

11. __________ hammer is used to reach the corners of bent sheets.

12. To strengthen the folded edge, __________ beading is done.

13. ______________ acid is used for cleaning the surfaces before soldering.

14. ______________ is used to set the folded joint.

15. A __________ is used to head a rivet.

16. ——————— stake can be used to make funnels.

LEADER

1. Appeals to the best in each person; problem-solver; advice-giver.
2. Thinks of ways to make people more productive; more focussed on goals.
3. Comfortable with people in their work places.
4. Arrives early; stays late.
5. Good listener.
6. Decisive.
7. Tolerant of open disagreement.
8. Has strong convictions.
9. Trusts people.
10. Wants anonymity for himself, publicity for his organisation.
11. Often takes the blame.
12. Gives credit to others.
13. Sees growth as by-product of search for excellence.
14. Openness.
15. Promotes from within.
16. Keeps promises.
17. Grooms leaders.
18. Sees mistake as learning.

NON-LEADER
(Follower, Manager, Boss, Supervisor)

1. Invisible; gives orders to staff, and excepts them to be carried out.
2. Thinks of personal rewards, status, and how he/she looks to outsiders.
3. Uncomfortable with people.
4. In late; usually leaves on time.
5. Good talker.
6. Uses committees, consultants.
7. Intolerant of open disagreement.
8. Vacillates when a decision is needed.
9. Trusts only words and numbers on paper.
10. The reverse.
11. Looks for a scapegoat.
12. Takes credit; complains about lack of good people.
13. Sees growth as primary goal.
14. Secrecy.
15. Always searching outside the organization.
16. Does not.
17. Inhibits growth of colleagues.
18. Sees mistakes as punishable offenses.

Quality Circles in Japan

Some of the values which the Japanese cherish, religiously uphold and meticulously practice, explain for their **value based** approach to **Quality Circles,** anchoring it into their cultural and economical strong hold in the world. They are:

(a) **SEIRI, SEITON, SEIKSTA:** These three **S**'s mean the virtues of **tidiness, orderliness** and **cleanliness** at all times and at every place. These account for the clinically clean shop floors, roads and also in their personal habits.

(b) **ICHIGAI:** This simply means **"Being Worthwhile".** Every Japanee wants to be worthwhile to his country, company (employer) and society. He wants to be worth in the estimate of the others. Every Japanee gives of his best, wherever he goes, no matter whether his job is of a president or on attender. The quality circles give the workmen opportunity to demonstrate their self - worthiness in ample measure.

(c) **SEMPAI - KOHAI:** These two words explain in a nutshell the **senior - Junior relationship** and account for the harmonious relationship between **Workers, Supervisors** and **Managers.** The senior looks upon his junior like a younger brother and develops him. The senior earns his respect from the junior through his role as teacher and guide.

(d) **TATEMAE - HONNE:** It means **"leader always in front"**. While taking the responsibility to lead, the leader gives the credit to the team for its success and believes in consensus and respect human values.

(e) **WAAH :** This means **"being in peace"** with oneself. Could there be a better way of being at peace than working as a small team to do a good job?

Exercise - 4.1

Aim : To make a rectangular tray.	**Ex. No. :**
	Date :

Sketch :

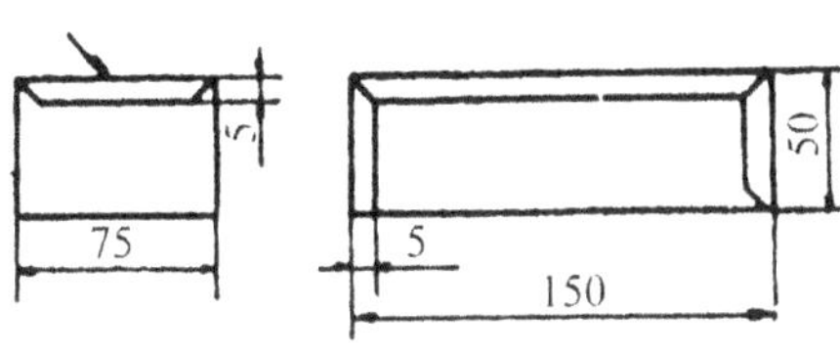

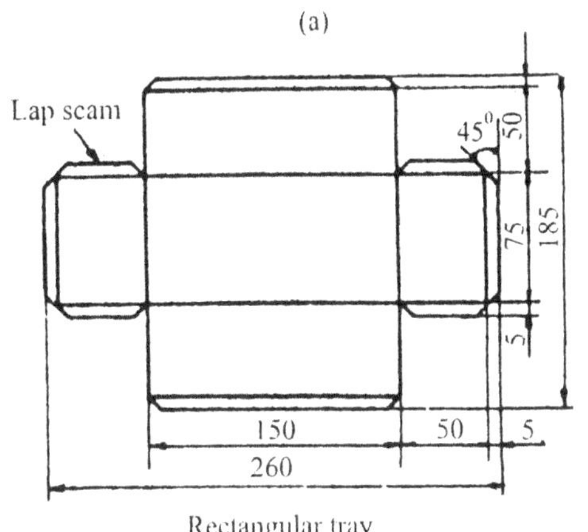

Rectangular tray

Material : Tin sheet

200 mm × 185 mm

List of tools used :

Sequence of operations :

Safety precautions :

Staff signature

Exercise - 4.2

Aim :	Ex. No. :
	Date :

Sketch :

Material :

List of tools used :

Sequence of operations :

Safety precautions :

Staff signature

Exercise - 4.3

Aim :	Ex. No. :
	Date :

Sketch :

Material :

List of tools used :

Sequence of operations :

Safety precautions :

Staff signature

Exercise - 4.4

Aim :	Ex. No. :
	Date :

Sketch :

Material :

List of tools used :

Sequence of operations :

Safety precautions :

Staff signature

CHAPTER - 5

WELDING SHOP

INTRODUCTION

Welding is the process of joining two pieces of metal. It results in a joint that is equivalent in composition and chracteristics of the matals joined. The various welding processes are: 1.Electric arc welding, 2.Gas welding, 3.Thermit welding, 4.Resistance welding and 5.Friction welding. However, only electric arc welding is discussed here. In this process, the work pieces are melted along a common edge, to their melting point and then a filler metal is introduced to form the joint on solidification. The materials to be welded must be free from rust, scale, oil or other impurities, so as to obtain a sound weld.

ARC WELDING

In arc welding, the heat required for joining the metals is obtained from an electric arc. Transformer or motor generator sets are used as arc welding machines. These machines supply high electric currents at low voltages and an electrode is used to produce the necessary arc. The coated electrode serves as the filler rod and arc melts the surfaces so that the metals to be joined are actually fused together. Figure shows the principle of arc welding using a transformer.

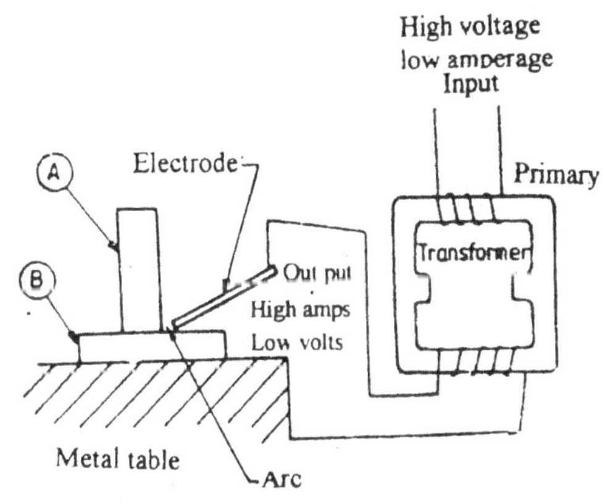

Arc welding

In addition to the welding machine, certain accessories are needed for carrying out the welding work.

Welding Cables

Two welding cables are required, one from the machine to the electode holder and the other, from the machine to the ground clamp. Flexible cables are usually preferred because of the ease of usage and coiling the cables. Cables are specified by their current carrying capacity, say 300A, 400A etc.

Electrodes

Filler rods used in arc welding are called electrodes. They are generally made of a rod of alloying elements suitable for the job, coated with a flux. They are specified by the diameter in SWG and length, apart from the brand and code names, indicating the purpose for which they are most suitable.

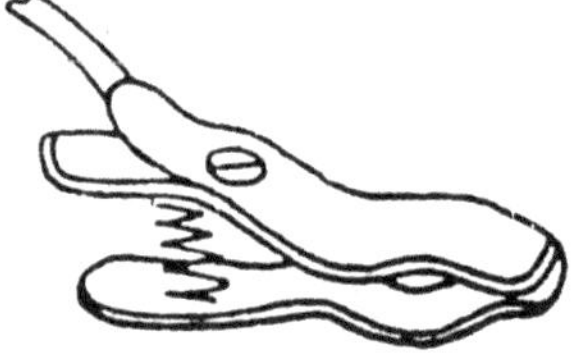

Electrode holder

Electrode Holder

The electrode holder is connected to the end of the welding cable and holdes the electrode. It should be light, strong and easy to handle and should not become hot while in operation. Figure shows one type of electric holder. The jaws of the holder are insulated, offering protection from electric shock.

Ground Clamp

It is connected to the end of the ground cable and is clamped to the work or welding table to complete the electric circuit It should be strong and durable and give low resistance connection.

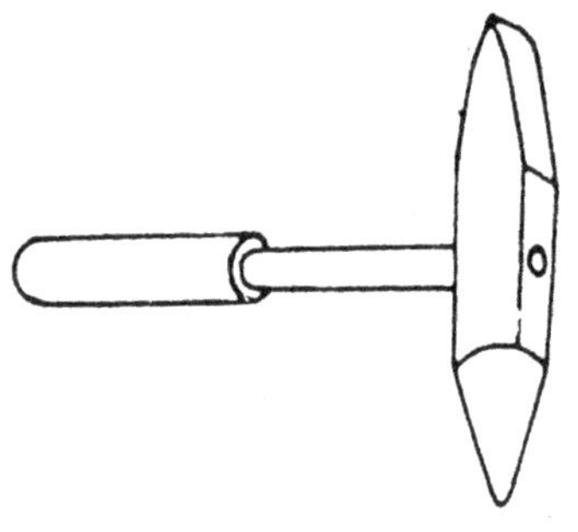

Ground clamp

Chipping Hammer and Wire Brush

A chipping hammer is used for removing slag formation on welds. One end of the head is sharpened like a cold chisel and the other, to a blunt, round point. A wire bursh is used for cleaning and preparing the work for welding.

Face Shield

A face shield is used to protect the eyes and face from the rays of the arc and from spatter or flying particles of hot metal. It is available either in hand or helmet type. The hand

Chipping hammer

type is convenient to use wherever the work can be done with one hand. The helmet type, though not comfortable to wear, leaves both hands free for the work.

There is nothing in the world that can not be improved upon.

Shields are made of light weight non-reflecting fibre and fitted with dark glass to filter out the harmful rays of the arc. A cover glass is fitted in front of the dark lens to protect it from spatter.

Weld Joints

Figure shows some common types of weld joints. Wherever possible, it is better to weld by placing the parts in the flat position.

PREPARATION OF WORK

Before welding, the work pieces must be thoroughly cleaned of rust, scale and other foreign materials. Thin pieces of metal are generally welded without bevelling the edges. However, thick work pieces should be bevelled to ensure adequate penetration and fusion of all parts of the weld. But, in either case, the parts to be welded must be separated slightly to allow better penetration of the weld.

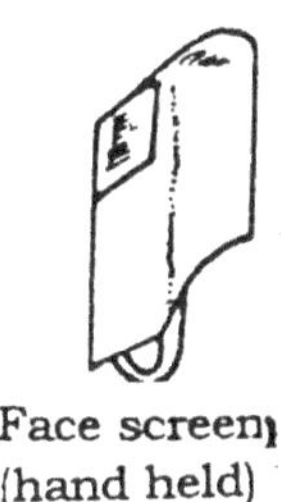

Face screen (hand held)

Face shield

Safety shields

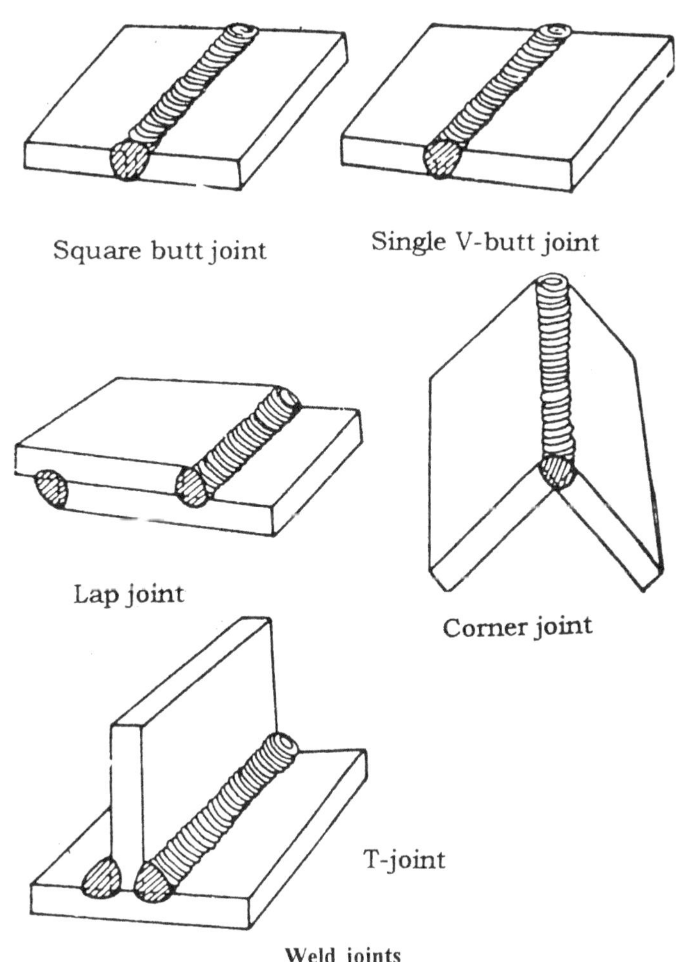

Square butt joint

Single V-butt joint

Lap joint

Corner joint

T-joint

Weld joints

The occupational disease of a poor executive is inability to listen.

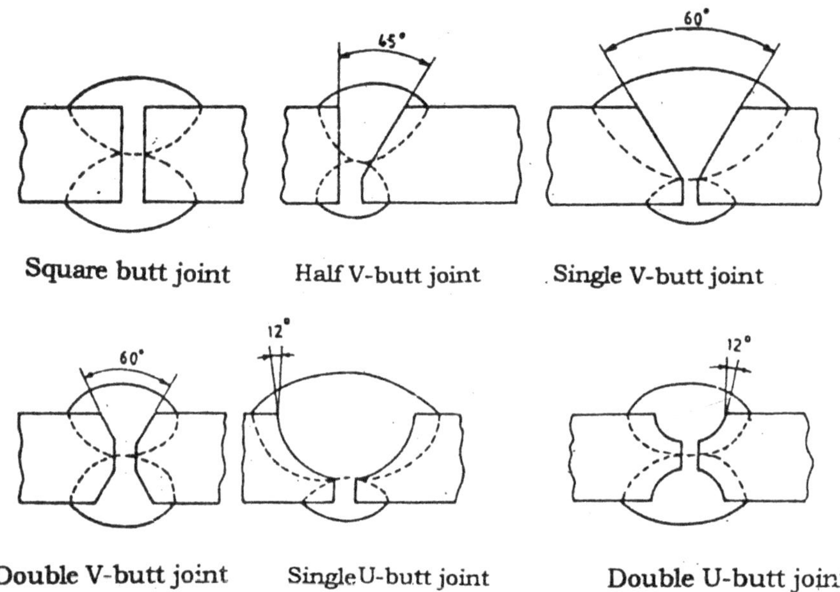

Square butt joint   Half V-butt joint   Single V-butt joint

Double V-butt joint   Single U-butt joint   **Double U-butt** joint

Edge preparation

WEAVING

A steady, uniform motion of the electrode produces a satisfactory bead. However, a slight weaving or oscillating motion is preferred, as this keeps the molten metal a little longer and allows the gas to escape, bringing the slag to the surface. Weaving also produces a wider bead with better penetrations as shown in figure.

EFFECT OF CURRENT AND SPEED

The current setting and the speed of electrode movement affect the penetration and the strength of the weld as shown in figure. In both the figures, weld (a) shows

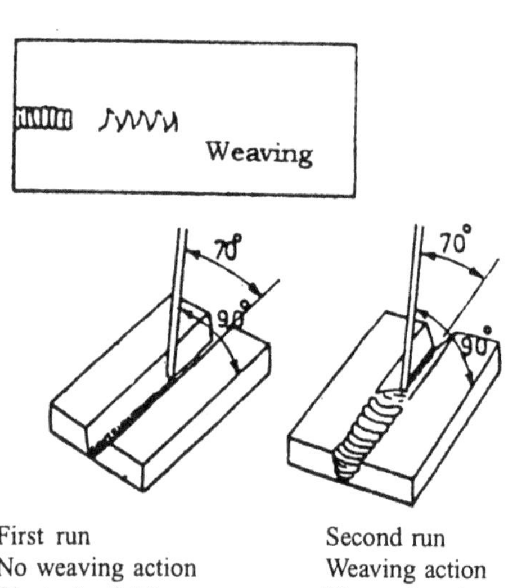

First run
No weaving action
Electrode: 3.15 mm
Current : 115A

Second run
Weaving action
Electrode : 4 mm
Current : 160 A

Single V-butt weld in 6 mm M.S plate

Weaving

The bottleneck is at the top of the bottle.

proper adjustment of current and electrode speed with good fusion and no overlap or undercutting. Too low a currrent results in high bead with poor penetration. Low speed results in large, wide bead with overlap, as shown at (b). Too high a current results in deep penetration, undercutting of the edges of the bead, and excessive spatter, whereas fast movement results in small, narrow, irregular bead with poor penetration as shown at (c).

WELDING POSITIONS

Depending on the location of the welding joint, appropriate position of the electrode and hand movement are selected. Figure shows different welding positions.

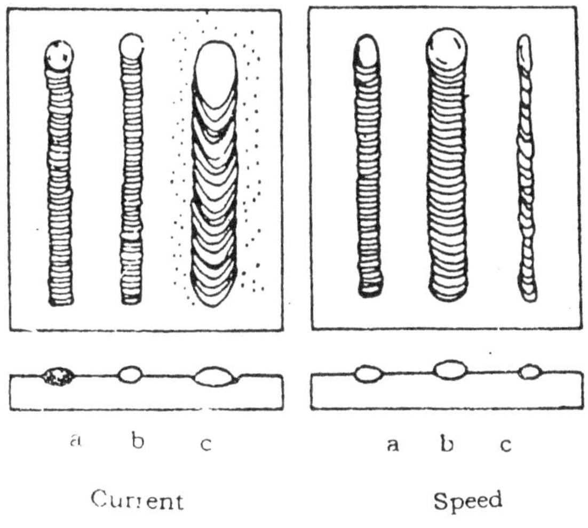

Effect of current and speed

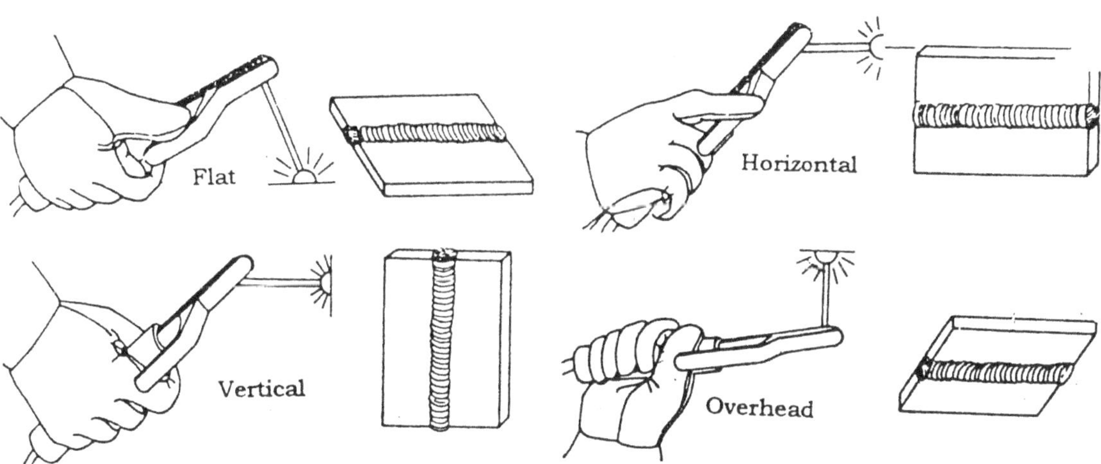

Welding Positions

Arc Welding Machines

Both direct current (DC) and alternating current (AC) are used for electric arc welding, each having its particular applications. DC welding supply is usually obtained from generators driven by electric motor or if no electricity is available by internal combustion engines.

Transformers are predominantly used for almost all welding works. They have to step down the usual supply voltage of 200 – 400 volts to the open circuit welding voltage of 50 – 90 volts. A 100A to 200A machine is small and portable and available in single phase also. A 300A or 400A size is suitable for manual welding of average work. Automatic welding requires 800 to 3000A.

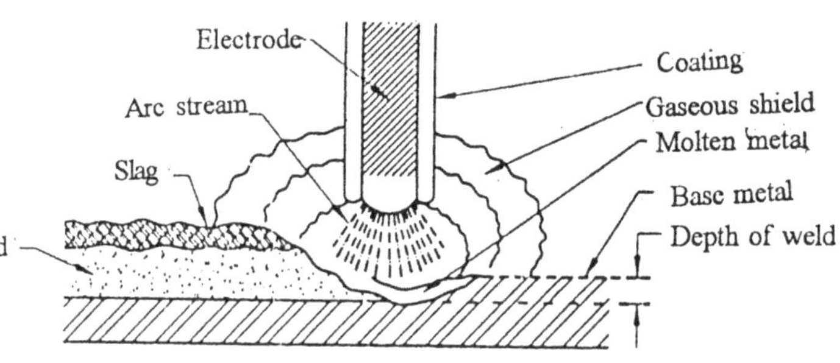

Arc welding

The two popular manufacturers are i). Indian Oxygen's INDARC and ii). Advani Orlikon makes. The transformer may be air cooled or oil cooled with transformer oil.

Arc Welding Currents and Voltages

The rage of currents and voltages applied are changed according to the electrode selected for the job. The details are given in tables below.

Table : Voltages used in arc welding

Current (A)	Voltage (V)
Upto 100	15
Over 100 to 250	20
Over 250 to 350	30
Over 350 to 500	35
Over 500	40

Table : Welding currents for various electrode sizes

Dia of Electrode mm	Current Amps
2	45
2.5	70
3.25	105
4.0	140
5.0	180
6.0	235
8.0	310

Coated electrodes

Electrodes may be plain without coating or with coating. The coating of electrodes is done to provide one or more of the purposes given below :

1. to facilitate the establishment and maintenance of the arc,
2. to protect the molten metal from the oxygen and nitrogen of the air by producing a shield of gas around the arc and weld pool;
3. to provide the formation of slag in order to protect the welding seam from rapid cooling, and
4. to provide a means of introducing alloying elements not contained in the core wire.

Polarity in arc welding

Because of the reversal of the current with AC supply, the heat generated at either pole is the same, and changing over the connections to the electrode does not have any effect on its performance. On the other hand, polarity of DC has a great bearing on electrode performance.

The conductor from which the direct current passes into the arc is designated as the positive terminal and the other conductor as the negative terminal. The heat generated by the flow of current is split into two parts in the ratio 66% at the positive pole and 33% at the negative pole. There are several ways of denoting the connection to the electrode when welding with DC. When the electrode is connected to the positive lead the term electrode positive is employed. It is also referred to as **reversed polarity**. Similarly, when the electrode is connected to the negative lead, the term electrode negative is employed. It is also referred to as **straight polarity**.

Advantages and disadvantages of AC and DC

The following are the advantages and disadvantages of AC and DC welding sets.

1. The advantage of DC welding lie in the higher arc stability and the degree to which the work is heated. Whilst direct current has the advantage of allowing a desired heat distribution, AC welding sets are presently gaining considerable ground.
2. The efficiency of AC welding transformer varies from 0.8 to 0.85, the efficiency of DC sets is very low and vary from 0.3 to 0.6.

3. The energy consumption per kg of deposited metal in AC welding is from 3 to 4 Kwh which for DC welding it is as high as 6 to 10 Kwh.

4. The disadvantage of AC is low power factor, which is 0.3 to 0.4 where as that of DC welding is 0.6 to 0.7.

5. In DC bare and thin coated electrodes can be used efficiently and polarity can be changed to suit the thickness of material to be welded. It is difficult to weld thin metals in AC when compared to DC.

6. When compared to AC transformer the initial cost of DC sets is two to three times more.

7. DC machines are suitable for non-ferrous metals.

Weld defects

In arc welding the defects commonly found are : i). lack of fusion, ii). lack of penetration, iii). inclusion of slag or oxide, iv). presence of cracks, v). porosity, vi). and under cut or excessive penetration (bad profile).

These defects are shown in figure below :

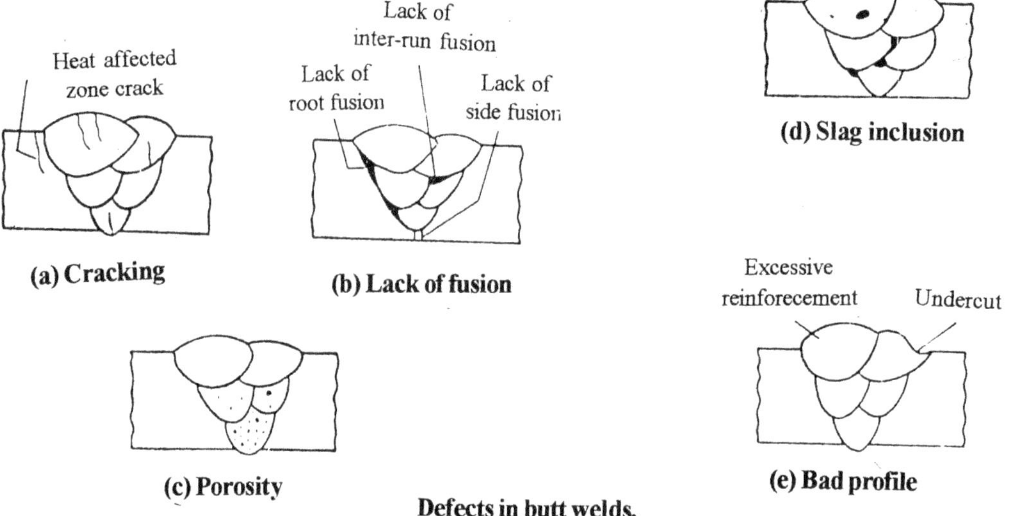

Defects in butt welds.

These defects have serious consequences if the material runs a risk of brittle fracture or the joint is subjected to a fatigue loading. The presence of a crack always enhances the probability of a brittle fracture. Similarly, a lack of fusion causes a sharp discontinuity which in turn reduces the fatigue strength.

Welding Symbols

Conventional representation of welding in drawings is shown in the table below.

DESIGNATION	ILLUSTRATION	SYMBOL
Square butt weld		||
Single V - butt weld		V
Single V - butt weld with broad root face		Y
Fillet weld		△

SKETCH OF WELD	REPRESENTATION	DESCRIPTION
		A. Fillet welds to be made on the arrow side of the joint. B. Fillet welds to be made on the otherside of the joint.
		Fillet welds to be made on either side of the joint.
		Fillet weld to be made all round
		Fillet weld to be made at site and size of fillet weld is to be 6mm

Conventional Representation of Welds

GAS WELDING

Oxyacetylene flame is commonly used for gas welding. It consists of the supply of oxygen and acetylene under pressure in cylinders, pressure regulators, a torch, hoses and accessories like goggles and a lighter. The oxygen and acetylene cylinders are connected to the torch through pressure regulators and hoses. The regulator consists of two pressure gauges, one for indicating the pressure within the cylinder and the other shows the pressure of the gas fed into the torch. which may e regulated. The torch mixes the two gases and the flame may be controlled by adjusting the oxygen and acetylene supply.

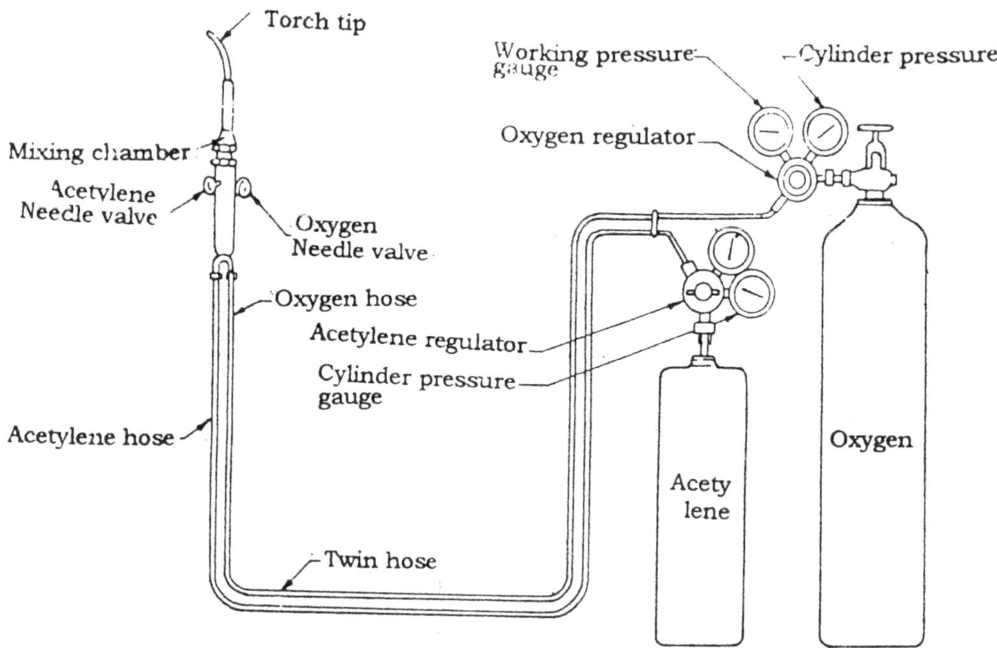

Gas welding equipment

Goggles

Goggles with coloured glasses are used to protect the eyes from glare and flying bits of hot metal. A welding table with a top of fire bricks is recommended for oxyacetylene welding.

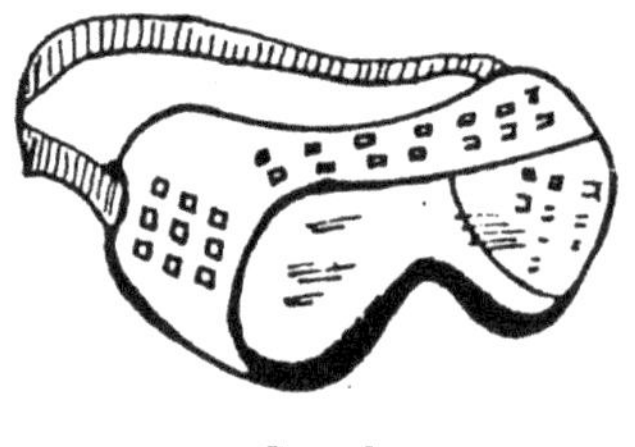

Goggles

Types of Flames

The correct adjustment of the flame is important for efficient welding. When oxygen and acetylene are supplied to the torch in nearly equal volumes, a neutral flame is produced having a maximum temperature of 3200°C. The neutral flame is widely used for welding steel, stainless steel, cast iron, copper, aluminium etc. Carburising flame produced with an excess of acetylene, is needed for welding lead, Oxidising flame with excess of oxygen is used for welding brass.

Depending on the thickness of the job, different torch nozzle sizes are used. The pressure of the gases and the flame size vary depending on the size of the nozzle tip.

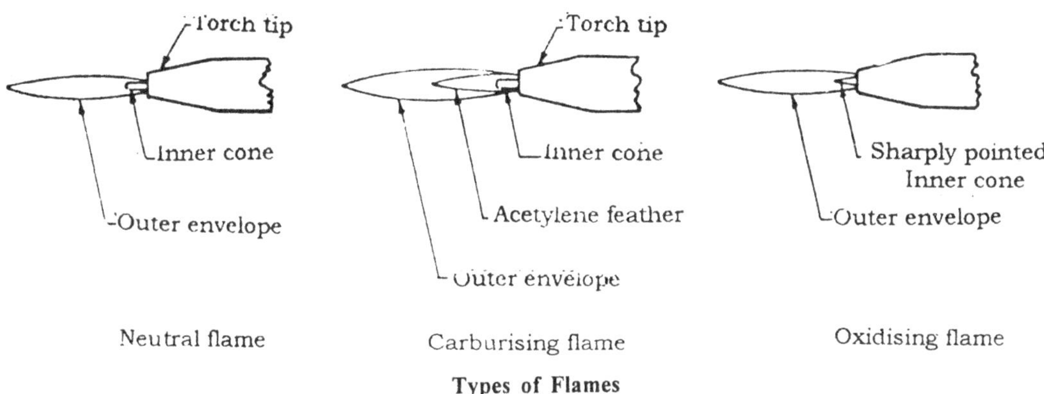

Types of Flames

Filler Rods

For oxyacetylene gas welding, filler rods are not coated with flux; however they are applied separately, Mild steel welding rods are usually copper coated to prevent rusting. Cast iron rods are square shaped. Brazing rods are made of brass or bronze. They are usually one metre long. Filler rod size increases as the metal thickness to be joined increases. 1.5 mm diameter filler rod is recommended for 18 SWG sheet and 2 to 3 mm diameter for 3 mm thick sheet and so on.

Type of Joints

The type of joint needed depends on the nature of material, its thickness and the kind of job. The types of joints used are common to both arc and gas welding. Both ferrous and non-ferrous welding may be carried out in gas welding.

Technique of welding

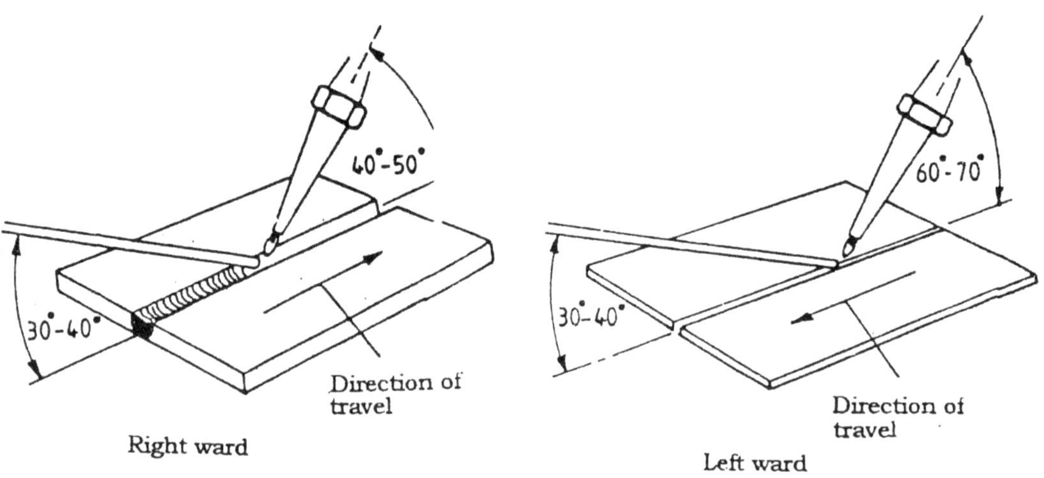

Technique of welding

Adjusting equipment and lighting the torch

The recommended stages of adjusting the gas welding equipment and lighting the torch are as follows:

1. Select the proper size tip for the job and insert it carefully into the torch.

2. Check the valves on the torch to ascertain that they are turned-off (clockwise).

3. Open the acetylene cylinder valve slightly, say ¼ to ½ turn.

4. Open the oxygen cylinder valve slowly, till it is fully open.

5. Open the acetylene valve on the torch and turn the acetylene regulator screw clockwise, until the gauge reads 0.5 to 1 kg/cm² of pressure. Then close the valve on the torch.

6. Open the oxygen valve on the torch to check the flow and close it,

7. Put-on the welding goggles, gloves and apron.

8. Open the acetylene valve on the torch by ¼ turn. Light the torch with a lighter, keeping its tip away from the cylinders and your body.

9. Adjust the acetylene valve on the torch until the flame extends slightly from the end of the tip.

10. Open and adjust the oxygen valve on the torch until the desired flame is obtained.

Welding process

The following are the steps involved in a gas welding work:

1. Prepare the work pieces to be welded and place them in proper position on the welding table.

2. Wear goggles, gloves and apron.

3. Select proper size tip for the job and fix it to the torch.

4. Select the filler rod of recommended size.

5. Adjust the welding equipment and light the torch.

6. Adjust the torch for neutral flame.

7. Hold the torch with the inner cone, about 3 mm away from the metal and tack-weld the pieces at either end.

8. Starting from one end, weld along the edge with a zig-zag torch movement. Add the filler metal to the joint as welding progresses. The two techniques of gas welding are shown in figure. Leftward is recommended for thin sheets and rightward for thick sheets.

Note: Do not touch the torch tip with the rod or molten metal. Always keep the rod in the melt. If the torch tip is too close to the melt, it will form small blow holes in the weld and the torch may back fire. Practice the rhythm of torch and rod movement for achieving good results.

Shutting-off the equipment

After completing gas welding operation, the following procedure must be followed for shutting-off the equipment:

1. First close the torch acetylene valve and then torch oxygen valve.

2. Close the acetylene cylinder valve first and then oxygen cylinder valve.

3. Drain the gas from the regulator and hose by opening the torch acetylene valve.

4. Drain the oxygen from the regulator and hose by opening the torch oxygen valve.

5. Open the regulator screws on each regulator and remove pressure from the diaphragms of the regulators.

6. Hang-up the hose and torch.

SAFE WELDING PRACTICES

Arc Welding

1. Never look at the arc with the naked eye. The are can burn your eyes severely. Always use a shield while welding.

2. Always wear the safety hand gloves, apron and leather shoes.

3. Ensure proper insulation of the cables and check for openings.

4. Apply eye drops after arc welding is over for the day to relieve the strain on the eyes.

Gas Welding

1. Always wear welding goggles while doing gas welding.

2. Never play or get careless when using the gas welding equipment. Combustible gas must be handled carefully.

3. Always use the spark lighter to light the torch, never use a match.

4. Strictly follow the procedures laid-out in handling the cylinders, regulators and the torches.

OBJECTIVE QUESTIONS

Fill in the blanks.

1. Welding is a process of joining metal pieces by ________________________________

2. Name three basic types of joints, common to both oxy-acetylene and arc welding.

3. After the arc has been obtained, the electrode is kept within ____________________ mm of the metal being welded.

4. ____________________ is used to remove the slag formed on the weldment.

5. What is the position of welding normally practiced?

6. What type of flame is best for most welding jobs?

7. If the torch tip gets too close to the metal being welded, it will cause small ____________________ to form in the metal.

8. What size of filler rod would you use when welding 18 gauge metal?

9. The torch mixes _________________ and _____________________ gases to provide fuel for the flame.

10. The maximum temperature obtained with neutral flame is _____________.

11. High current rating results in _________________ defect on the job.

12. Low speed of electrode movement results in _____________ bead.

13. The current rating recommended for welding 6mm M.S. plates with 10 SWG electrode is _____________________ amps.

14. Both _________________ pressure and _________________ pressure welding torches are available.

15. Acetylene gas is stored in the cylinders in the _________________ state.

16. Mild steel welding rods are coated with _________________, to prevent rusting.

17. _________________ flame is used for welding brass.

18. The movement of welding rod for better bead is known as _______________ .

19. _________________ bricks are used on the table for gas welding.

20. Filled acytelene gas cylinders are always kept_________________.

Safety Precautions in Welding:

Welding on colsed inflamable container
results in dangerous explosions

Seeing the electric are without the
shield will burn your eyes.

Exercise - 5.1

Aim : To make a square butt weld.	**Ex. No. :**
	Date :

Sketch :

Square butt joint

Material : MS flat

32 mm × 75 mm 5 or 3 mm

List of tools and equipment :

Sequence of operations :

Safety precautions :

Staff signature

Exercise - 5.2

Aim :	Ex. No. :
	Date :

Sketch :

Material :

List of tools and equipment :

Sequence of operations :

Safety precautions :

Staff signature

Exercise - 5.3

Aim :	Ex. No. :
	Date :

Sketch :

Material :

List of tools and equipment :

Sequence of operations :

Safety precautions :

Staff signature

Exercise - 5.4

Aim :	Ex. No. :
	Date :

Sketch :

Material :

List of tools and equipment :

Sequence of operations :

Safety precautions :

Staff signature

CHAPTER - 6

BLACKSMITHY SHOP

INTRODUCTION

Blacksmithy work consists of heating a metal stock till it acquires sufficient plasticity, followed by hand forging involving hammering, bending, pressing etc., till the desired shape is attained. Hand forging is the term used, when the process is carried out by hand tools. The hand forging process is generally employed for relatively small components.

TOOLS

Forge or Hearth

A smith's forge or hearth is used to heat the metal to be shaped. Hearths are used for heating small jobs to be forged by hand. Gas, oil or coal firing may be used for the purpose. The required air for the fire is supplied under pressure, by a blower through the tuyere in the hearth. The blowers may either be hand operated or power driven. In the latter case, the amount of air supply is controlled by valves near the forge. The hood, collects gases of combustion and sends through a chimney. The following are the forging temperatures of various metals:

Table 1 Forging temperature of metals

Metal	Forging Temperature $^\circ$C
Mild steel	750-1300
Wrought Iron	900-1300
Medium Carbon Steel	750-1250
High carbon and Alloy steel	800-1150

Anvil

It provides the necessary support during forging by resisting the heavy blows rendered to the job. It is also useful for operations such as bending, swaging etc. Its body is generally made of cast steel, Wrought iron or mild steel, with a hardened top layer of about 20 to 25 mm thick. Figure shows an anvil with various parts marked on it.

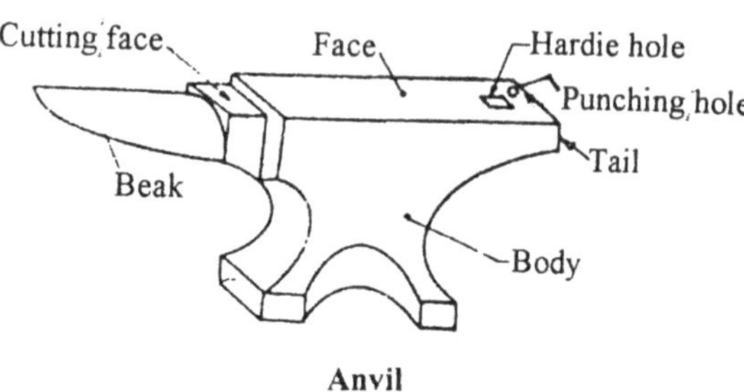

Anvil

Beak is used for bending metal to round shapes. The hardie hole which is square, is used to hold square shank tools like hardies, swages and fullers. The punching hole is used for bending small rods and punching holes in the work. Anvils are made in sizes weighing from 25 to 250kg. An anvil weighing about 75kg is suitable for general purpose.

Swage Block

A Swage block has a number of slots of different shapes and sizes along its four side faces and through holes of different shapes and sizes, running from its top to bottom faces. This is used as a support while forming (Swaging) different shapes and in punching holes. It is generally made of cast iron or cast steel.

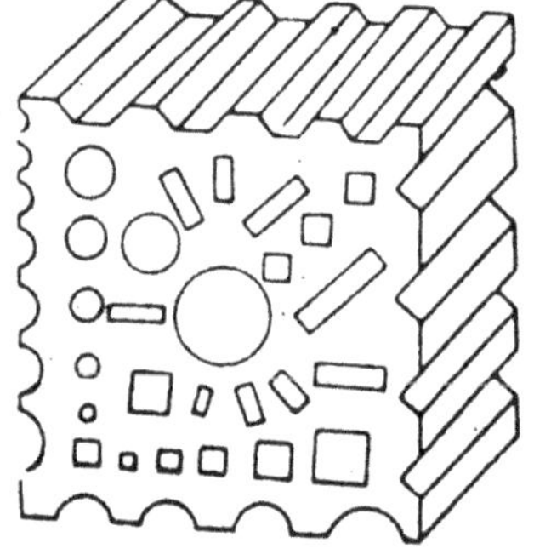

Swage block

Leg Vice

It is a heavy duty vice, fixed to the work bench at one end of a leg or set in a concrete base. It is mainly used for light forging and bending work.

Hammers

Hammers of different types and weights are used in blacksmithy. The ball peen hammer used for forging, weighs 0.5 to 1kg. The sledge hammer which is used for heavy work, has flat ends on either side and weighs 3 to 8 kg.

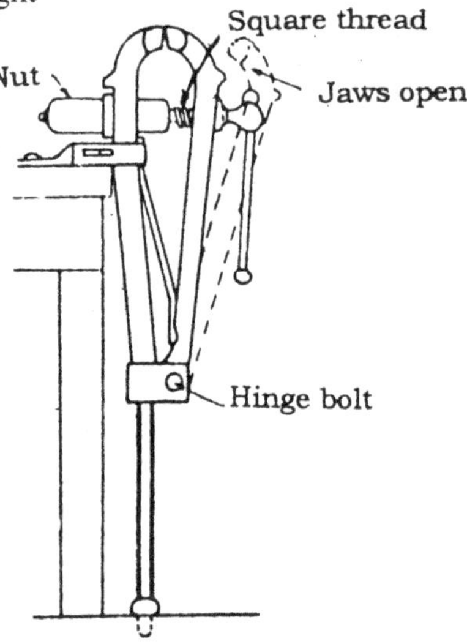

Leg vice

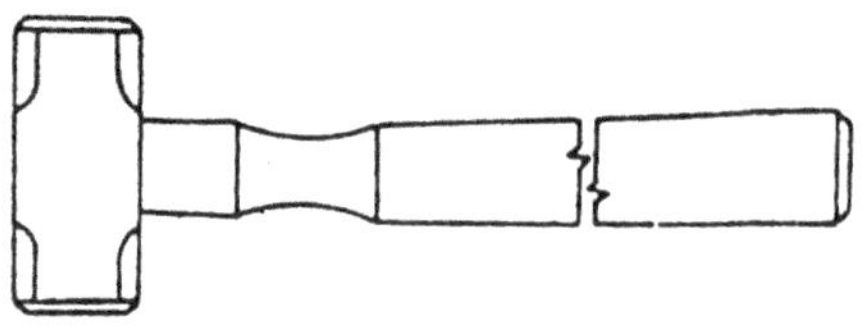

Sledge hammer

The length of the handle of a hammer increases with its weight.

Tongs

The metal to be forged must be held securely, while it is being shaped. A pair of tongs of suitable size and shape must be used for the purpose. Figure shows the most commonly used shapes in a blacksmithy shop. They are made of mild steel and the sizes vary from 40 cm to 60cm in length and 6mm to 55mm opening.

FORGING OPERATIONS

The following are the basic operations that may be performed by hand forging.

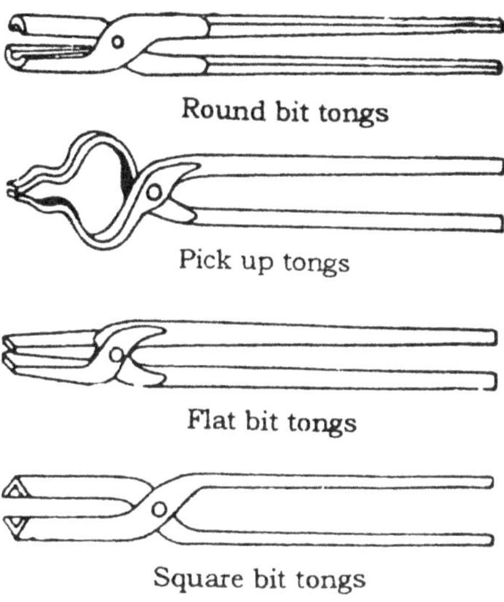

Tongs

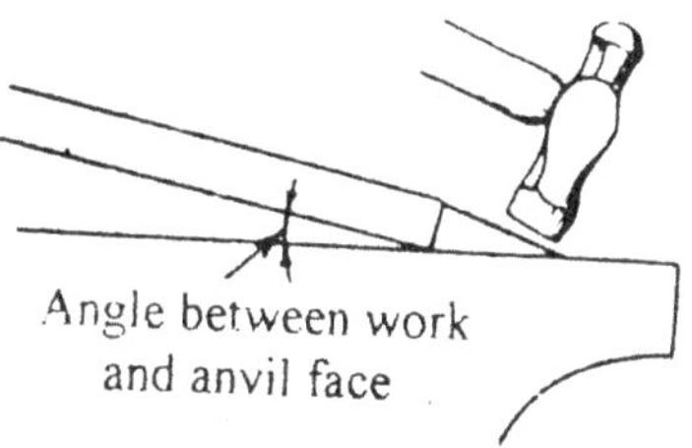

(a) Drawing down to short point

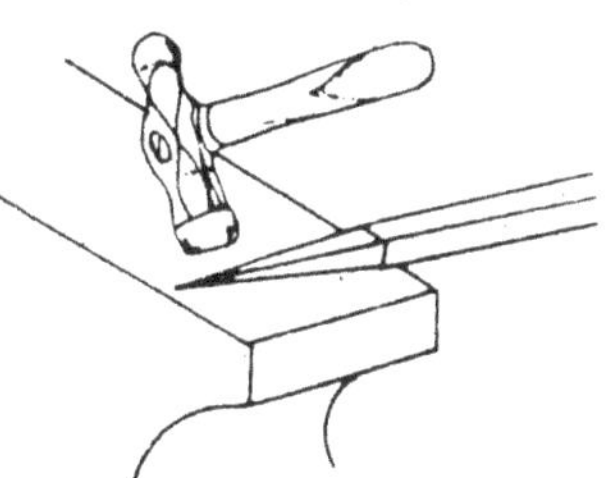

(b) Drawing down to a long square point

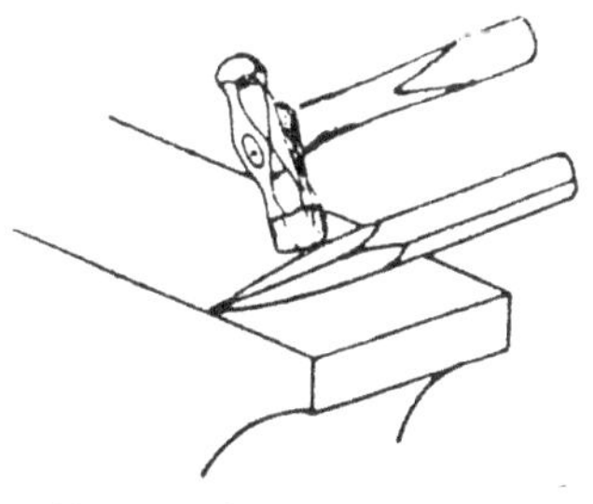

(c) Hammering down the corners to change the point to octagon

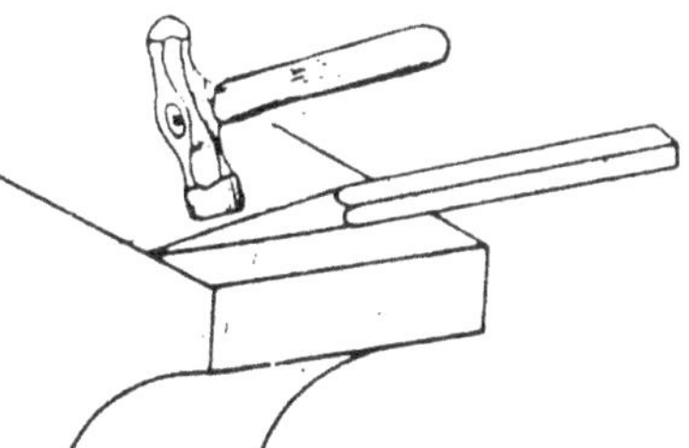

(d) Drawing down from octagonal to round

Drawing

Drawing

Drawing is the process of stretching the stock while reducing its cross section. Forging the tapered end of a cold chisel is an example of drawing operation. The steps in drawing down to a point are shown in figure.

Upsetting

It is a process of increasing the area of cross section of a metal piece locally, with a corresponding reduction in length. In this, only the portion to be upset is heated to forging temperature and the work is then struck at the end with a hammer as shown in figure. Hammering is done by the smith (student) himself, if the job is small or by his helper, in case of big jobs when heavy blows are required with a sledge hammer.

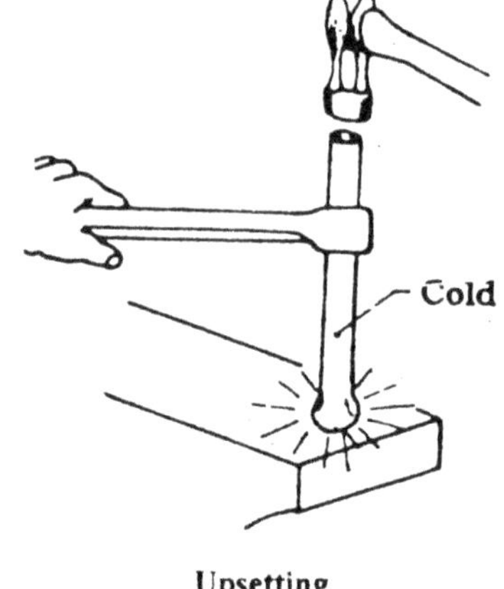

Upsetting

Fullering

Fullers are used for necking down a piece of work, the reduction often serving as the starting point for drawing. Fullers are made of high carbon steel in two parts, called the top and bottom fullers. The bottom tool fits in the hardie hole of the anvil. Fuller size denotes the width of the fuller edge.

Flattening

Flatters are the tools that are made with a perfectly, flat face of about 7.5 cm square. These are used for finishing flat surfaces. A flatter of small size is known as set-hammer and is used for finishing near corners and in confined spaces.

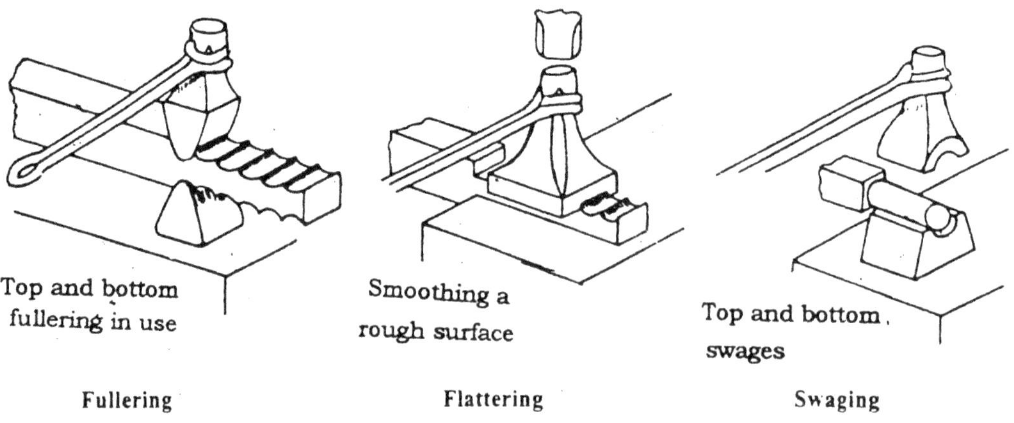

Fullering	Flattering	Swaging

Curiosity and experimentation have made men rich.

Swaging

Swages, like fullers are also made of high carbon steel and are made in two parts called the top and bottom swages. These are used to reduce and finish to round or hexagonal form. For this, the swages are made with half grooves of dimensions to suit the work.

Bending

Bending of bars, flats etc., is done to produce different types of bent shapes such as angles, ovals, circles etc. sharp bends as well as round bends may be made on the anvil, by choosing the appropriate place on it for the purpose. The stages in bending a loop is shown in above.

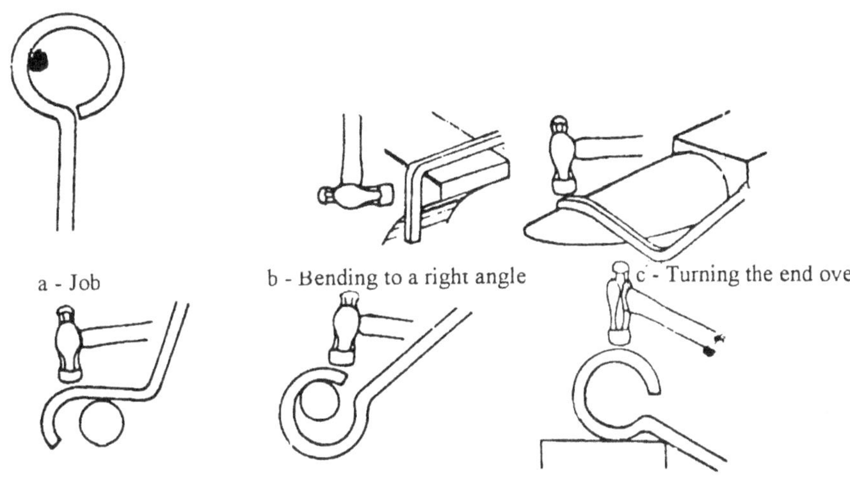

Bending

Twisting

It is also one form of bending. Sometimes, it is done to increase the rigidity of the work piece. small pieces may be twisted by heating and clamping a pair of tongs on each end of the section to be twisted and applying a turning moment. Larger pieces may be clamped in a leg vice and twisted with a pair of tongs or a monkey wrench. However, for uniform twist, it must be noted that the complete twisting operation must be performed in one heating.

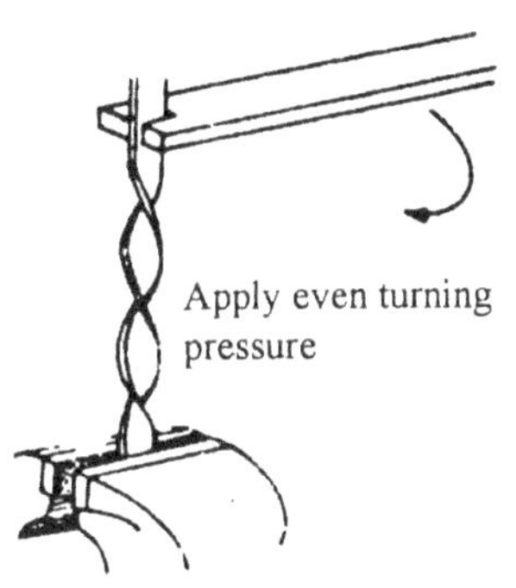

Twisting

Cutting (Hot and cold chisels)

Chisels are used to cut metals, either in hot or cold state. The cold chisel is similar to fitter's chisel, except that it is longer and has a handle. A hot chisel is used for cutting hot metals and its cutting edge is long and slender when compared to cold chisel. These chisels are made of tool steel, hardened and tempered. Figure shows the two chisels in operation.

Hot and cold chisels

Iron and Carbon Alloy

If the carbon content is less than 2 per cent in the iron-carbon alloy, it is known as steel. Again, based on the carbon content, it is called-mild steel. medium carbon steel and high carbon steel. The heat treatment to be given to these steels and their applications are shown in Table 2.

Table 2 Plain carbon tool steels

	Carbon %	Hardening temp °C	Tempering temp °C	Applications
Mild steel	0.1 0.25	800-840	250-300	Chain, rivets, softwire, sheet, tube, rod, strip
	0.5		"	Girders
	0.1 0.25	800-840	250-4300	Chain, rivets, softwire, sheet, tube, rod, strip
	0.5		"	Girders
	0.9		"	Taps, dies, punches, hot shearing blades
	1.0		"	Drills,reamers, cutters, blanking and slotting tools large turning tools
High carbon steel	1.2		"	Small cutters, lathe and engraving tools, files, drills
	1.35	720-760	"	Extra hard, planing, turning and slotting tools, dies and mandrels
	1.5		"	Razor blades

Happiness depends on what you can give, not on what you can get.

SAFE PRACTICES

1. Use correct size of tongs when forging short work; otherwise the job will fly, causing injuries.
2. Hold the hot work, downwards close to the ground, while transferring from the forge to anvil, to minimise danger of bums; resulting from accidental collisions with others.
3. Keep stray forgings off the floor.
4. Wear face shield when hammering hot metal.
5. Wear gloves when handling hot metal.
6. Wear steel-toed shoes.
7. Protect the eyes from infrared rays from the hot object and from the fumace.
8. All hammers should be fitted with tight and wedged handles.

MODEL RECORD SHEET

Aim : To make a hexagonal headed bolt from a given round rod.

Tools required : Brass rule, 500 gm ball peen hammer. 1kg ball peen hammer, hexagon; bottom swage, flatter, half round tongs and pick-up tongs.

Sequence of Operations

1. Heat one end of the bar for sufficient length to make the head and jump as shown in figure.
2. Flatten the head by hammering against the end of a bush or steel plate, placed directly over the hardie hole in the anvil, through which the stock passes as shown in figure.
3. Swage the head ot size as shown in figure by using the hexagonal bottom swage.
4. Swage the round portion of the bolt.
5. Chamfer the head.

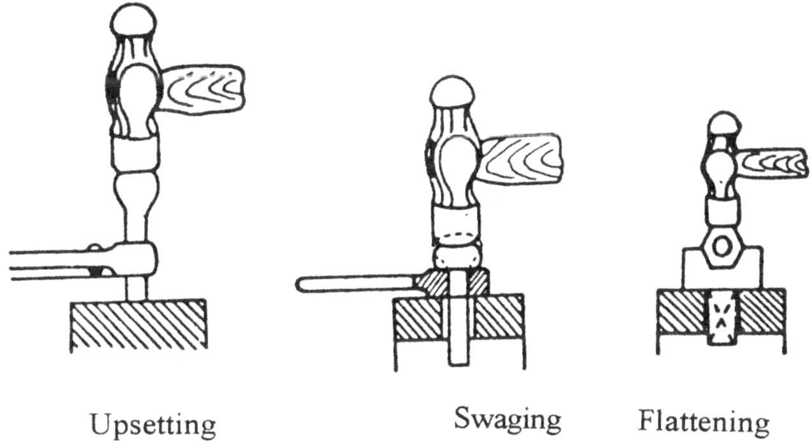

Forging a hexagonal bolt

MODELS FOR PRACTICE

1. Make a round bar from a square one. 2. Make a ring of square section from the given round rod. 3. Forge a crane hook. 4. Forge a 'S'hook from a given round rod. 5. Forge a 'T' headed bolt. 6. Forge a square headed bolt. 7. Forge a gib head sunk key. 8. Forge a chisel.

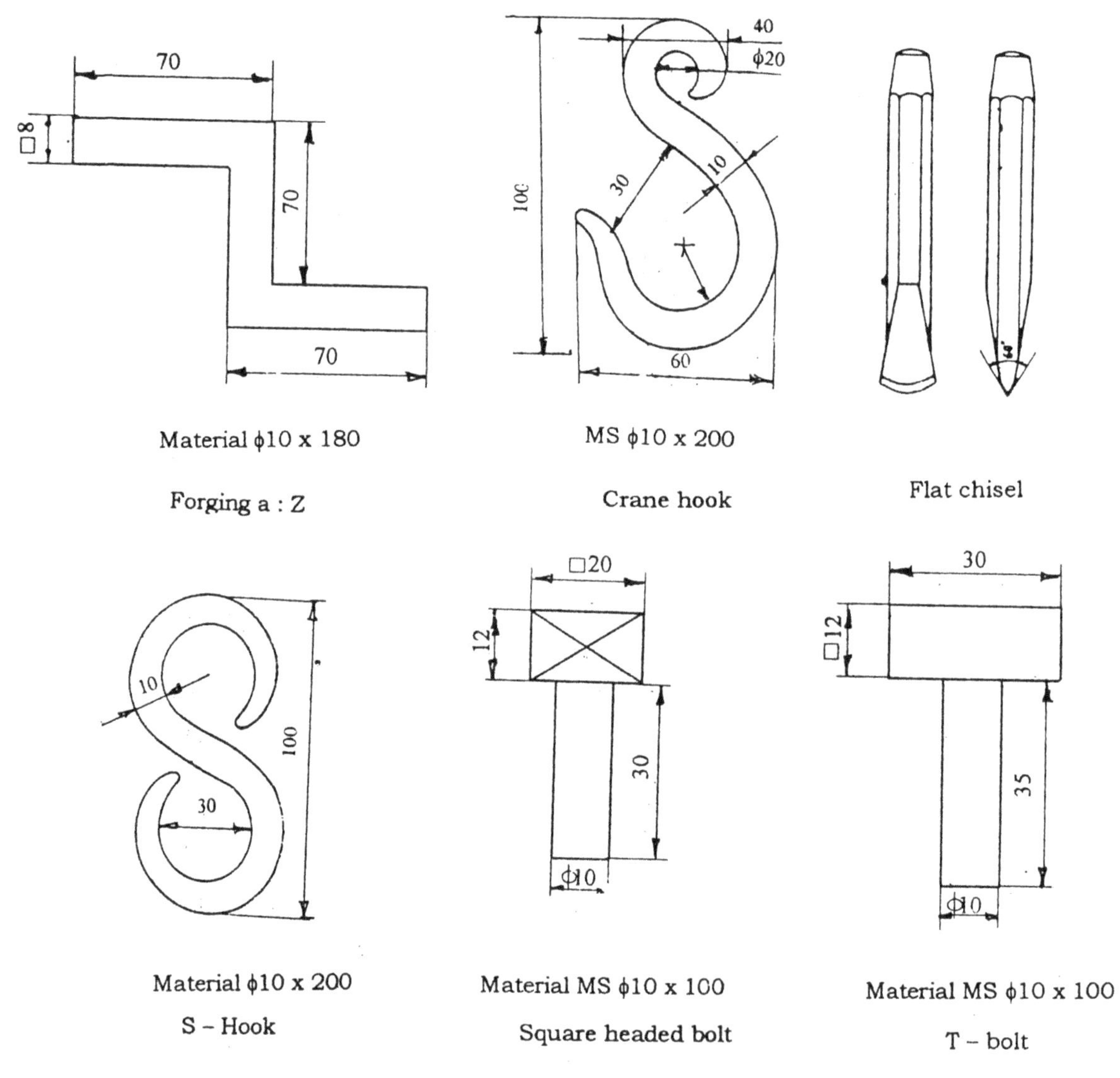

Models for practice

Manipulate every situation to give your best to mankind.

Exercise - 6.1

Aim : To foregea "Z"	**Ex. No. :**
	Date :

Sketch :

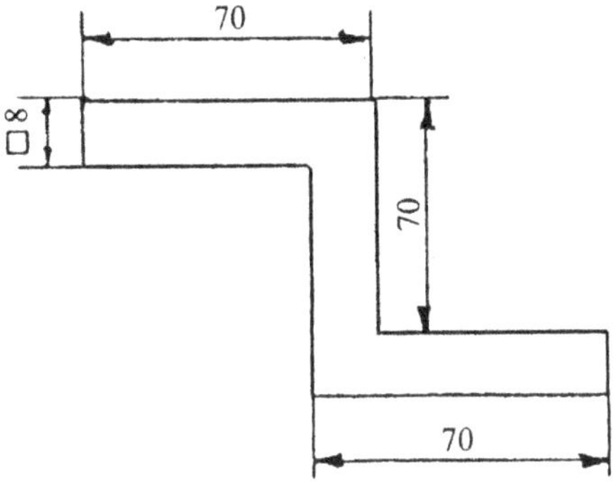

Material : Mild steel bar :

ϕ 10 × 180 mm

List of tools used :

Sequence of operations :

Safety precautions :

Staff signature

Exercise - 6.2

Aim :	Ex. No. :
	Date :

Sketch :

Material :

List of tools used :

Sequence of operations :

Safety precautions :

Staff signature

Exercise - 6.3

Aim :	Ex. No. :
	Date :

Sketch :

Material :

List of tools used :

Sequence of operations :

Safety precautions :

Staff signature

Exercise - 6.4

Aim :	Ex. No. :
	Date :

Sketch :

Material :

List of tools used :

Sequence of operations :

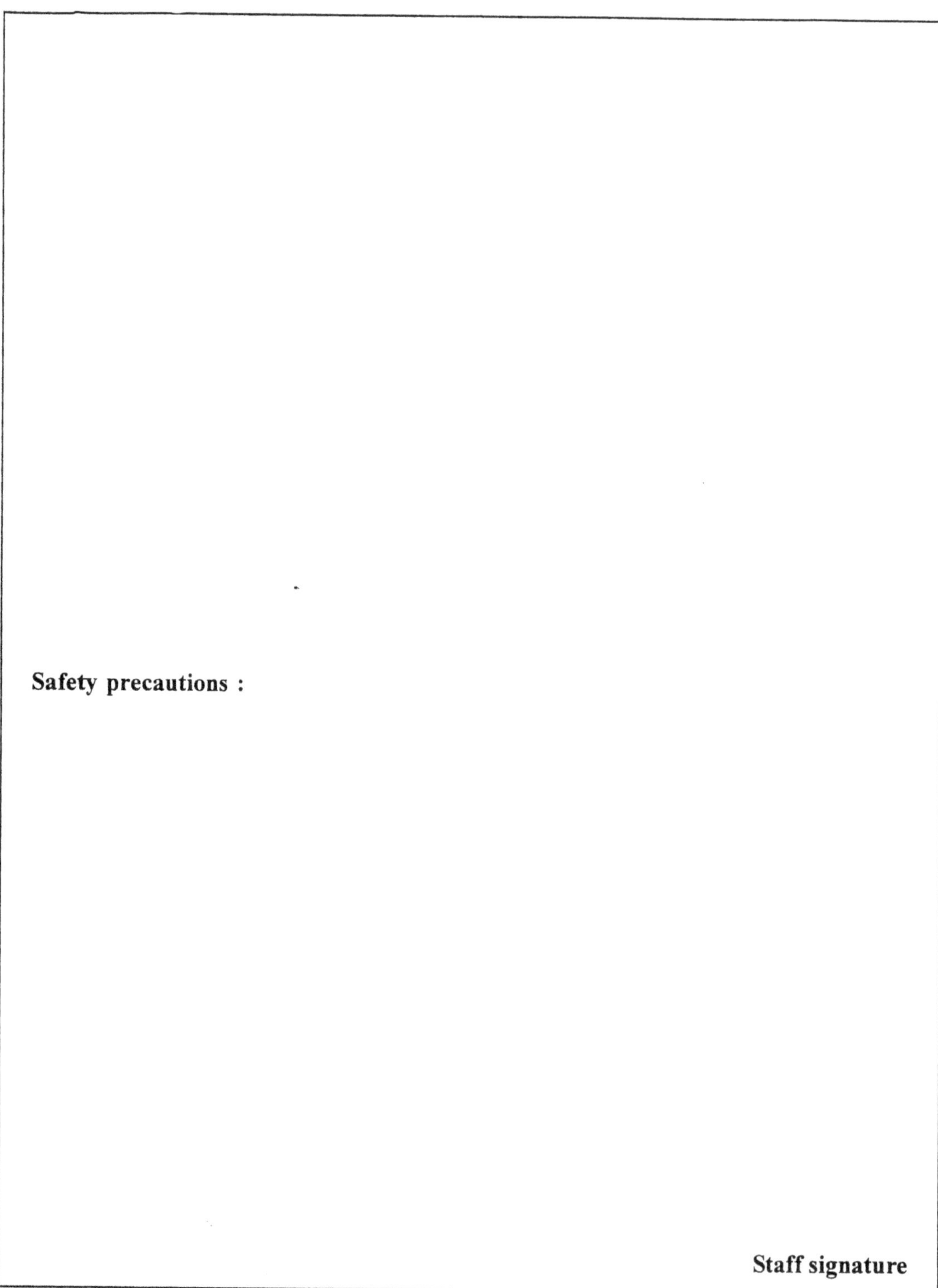

Safety precautions :

Staff signature

CHAPTER - 7

FOUNDRY SHOP

INTRODUCTION

Foundry practice deals with the process of making castings in moulds, formed in either sand or some other material. This is found to be the cheapest method of metal shaping. Further, castings may be made to fairly close dimensional tolerances by choosing proper moulding and casting process. In this chapter, the methods of making green sand moulds only are presented.

MOULDING SAND

Sand is the principal material used in a foundry. The principal ingredients of moulding sands are (i) silica sand, (ii) clay and (iii) water. Clay imparts the necessary bonding strength to the moulding sand. Moisture when added in correct proportion, provides the bonding action to the clay. Special additives and binders are also added to develop certain desired properties to the moulding sands. Natural moulding sand is either available in rever beds or dug from pits. It posseses an appreciable amount of clay and are used as received, with the addition of water. Synthetic sands are prepared by adding clay, water and other materials to silica sand, so that the desired strength and bonding properties are achieved.

Most of the moulding is done with green sand i.e, sand containing 6 to 8 percent moisture and 10 per cent clay content to give it sufficient bond. Green sand moulds are used for pouring the molten metal,immediately after preparing the moulds. Green sand moulds are cheaper and take less time to prepare. These are used for small and medium size castings. Dry sand moulds, obtained after drying or baking green sand moulds are used for large castings. Parting sand, which is clay free, fine grained silica sand, is used to keep the green sand from sticking to the pattern and also to prevent the cope and drags from clinging. Core sand is used for making cores. This is silica sand mixed with core oil and other additives.

PATTERNS

A pattern is the replica of the desired casting, which when packed in a suitable material, produces a cavity called the mould. This cavity when filled with molten metal, produces the desired casting after solidification.

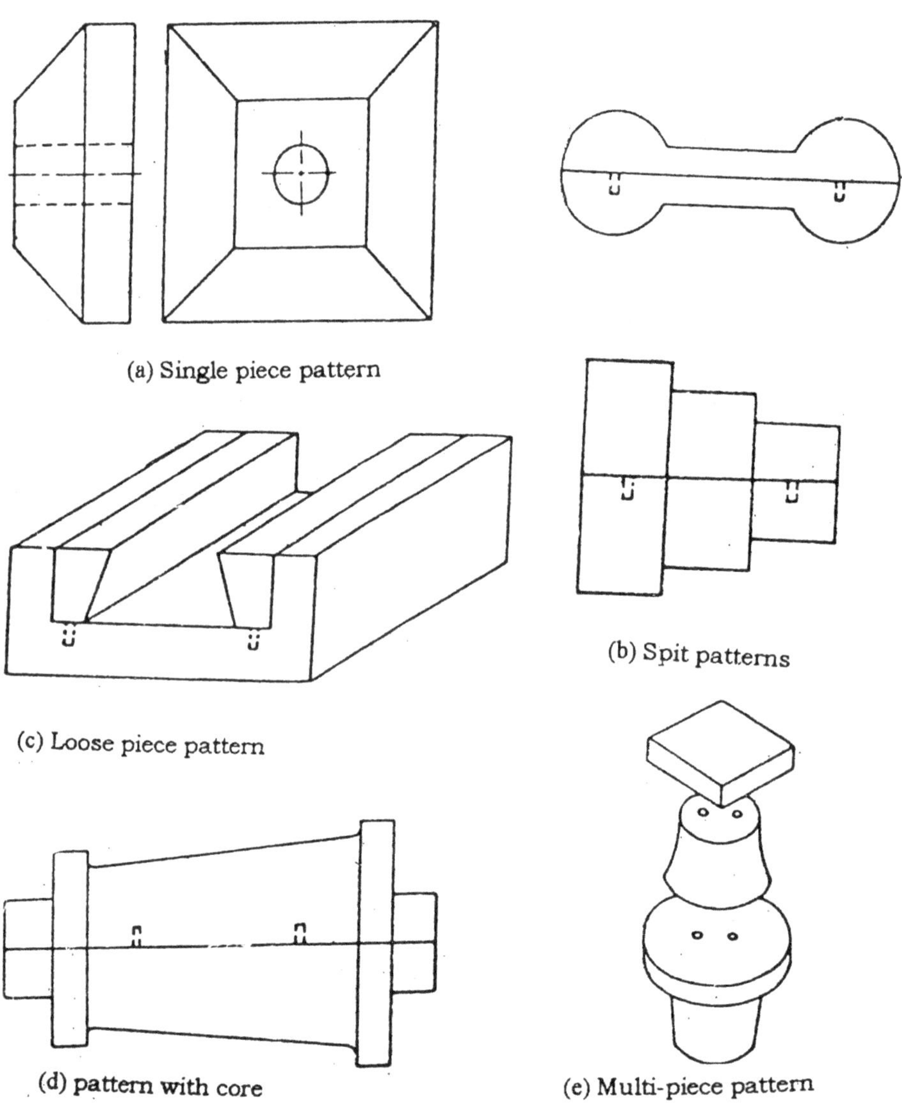

(a) Single piece pattern

(b) Spit patterns

(c) Loose piece pattern

(d) pattern with core

(e) Multi-piece pattern

Types of patterns

Types of Patterns

Wood or metal patterns are used in foundry practice. Single piece, split, loose piece and cored patterns are some of the common types.

Pattern Design

While designing a pattern, the following must be considered:

1. Avoid abrupt changes in cross section.

2. Avoid sharp corners and edges, to enable smooth flow of molten metal.

3. Provide the following pattern allowances.

i) Shrinkage allowance, to allow for shrinkage when casting cools in the mould.

ii) Slight taper or draft, to allow easy withdrawal of the pattern from the mould.

iii) Machining allowance, to take care of the machining on these surfaces.

TOOLS

The tools and equipment needed for moulding are: Moulding board, moulding flasks (boxes), shovel, and moulder's tools.

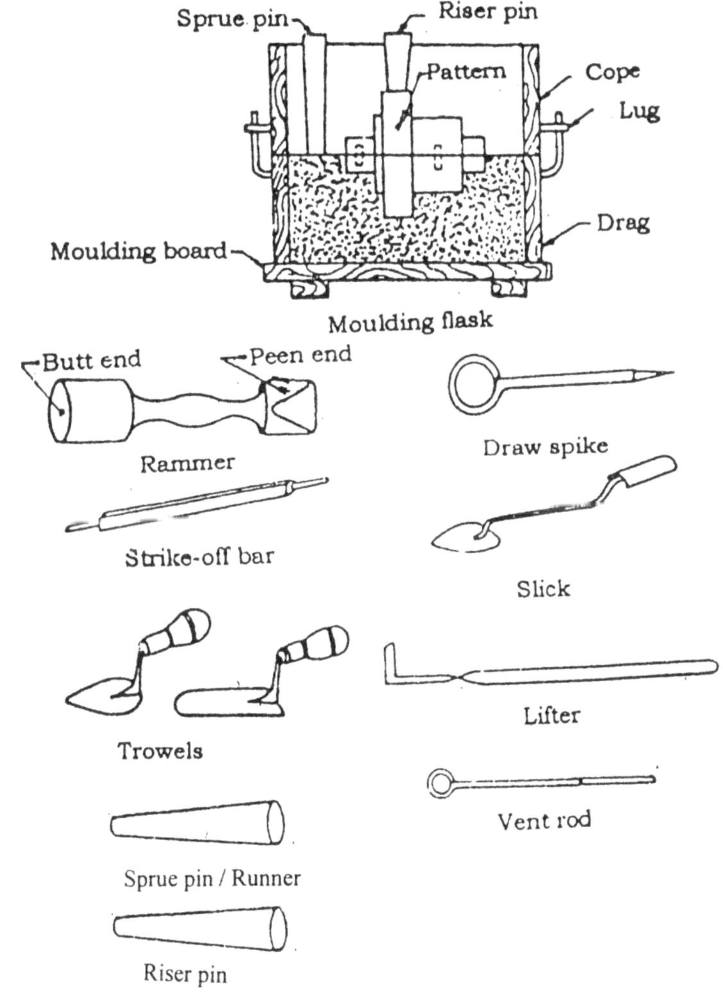

The mind is its own place, and in itself can make a heaven of Hell, a hell of Heaven.

Moulding Board

It is a wooden board with smooth surface. It supports the flasks and the pattern, while the mould is being made.

Moulding Flask

It is a box, made of wood or metal, open at both ends. The sand is rammed-in after placing the pattern to produce a mould. Usually, it is made of two parts. Cope is the top half of the flask, having guides for the aligning pins to enter. Drag is the bottom half of the flask, having aligning pins.

Shovel

It is used for mixing and tempering moulding sand and for transfering the sand into the flask. It is made of steel blade with a wooden handle.

Rammer

It is used for packing or ramming the sand around the pattern. One of its ends, called the peen end, is wedge shaped and is used for packing sand in spaces, pockets and corners, in the early stages of ramming. The other end, called the butt end, has a flat surface and is used for compacting the sand towards the end of moulding.

Strike off Edge

It is a piece of metal or wood with straight edge. It is used to remove the excess sand from the mould after ramming, to provide a level surface.

Sprue Pin

It is a tapered wooden pin used to make a hole in the cope sand through which the molten metal is poured into the mould.

Riser Pin

It is a straight wooden pin used to make a hole in the cope sand, over the mould cavity for the molten metal to rise and feed the casting to compensate the shrinkage that take place during solidification.

Trowel

It is used to smoothen the surface of the mould. It may also be used for repairing the damaged portion of the mould. Trowels are made in many different styles and sizes, each one suitable for a particular job.

Spike or Draw Pin

It is a steel rod with a loop at the other end. It is used to remove the pattern from the mould. A draw screw, with a threaded end may also be used for the purpose to draw metal patterns.

Slik

It is a small double ended tool having a flat on one end and a spoon on the other. It is used for mending and finishing small surfaces of the mould.

Lifters

Lifters are made of thin sections of steel of various widths and lengths, with one end bent at right angles. These are used for cleaning and finishing the bottom and sides of the deep and narrow pockets of the mould.

Gate Cutter

It is a semi-circular piece of tin sheet, used to cut gates in the mould. Gates are meant for easy flow of molten metal into the mould.

Bellows

It is a hand tool, used to blow air to remove the loose sand particles from the mould cavity.

Vent Rod

It is a thin rod used for making vents or holes in the sand mould to allow the escape of mould gases generated during the pouring of molten metal.

CORE BOX

A core box is desiged to mould cores. It is made of either wood or metal, into which core sand is packed to form the core. Wood is commonly used for making a core box, but metal boxes are used when cores are to be made in large numbers. Specially prepared core sand is used in making cores.

PREPARATION OF MOULD

The following are the steps used in making a simple mould using a single piece pattern.

(**Note:** Student is required to write the record in past tense, as done by him).

1. Place the pattern on the moulding board, with its flat side on the board.
2. Place the drag over the board, after giving a clay wash inside.

Never waste a minute thinking about people we don't like.

3. Sprinkle the pattern and the moulding board, with parting sand.

4. Allow loose sand, preferably throught a riddle over the pattern, until it is covered to a depth of 2 to 3 cm.

5. Pack the moulding sand around the pattern and into the corners of the flask, with fingers.

6. Place some more sand in the flask and pack the pattern with a rammer, using first the peen end and then butt end.

7. Strike-off the excess sand from the top surface of the drag with the strike-off bar.

8. Turn the drag upside down.

9. Blow-off the loose sand particles with the bellows and smoothen the upper surface.

10. Place the cope on top of the drag in position. Locate riser pin on the highest point of the pattern.

11. Place the sprue pin at about 5 to 6 cm from the pattern on the other side of the riser pin.

12. Sprinkle the upper surface with parting sand.

13. Repeat steps 3 to 7, appropriately.

14. Make holes with the vent rod to about 1 cm from the pattern.

15. Remove the sprue and riser pins by carefully drawing the mout. Make a funnel shaped hole at the top of sprue hole, called the pouring cup.

16. Lift the cope and place it aside on its edge.

17. Insert the draw pin into the pattern. Wet the edges around the pattern. Loosen the pattern by rapping. Then draw the pattern straight up.

18. Adjust and repair the mould by adding bits of sand, if necessary.

19. Cut gate in the drag from the spure to the mould, Blow-off any loose sand particles in the mould.

20. Close the mould by replacing the cope and placing weights on it.

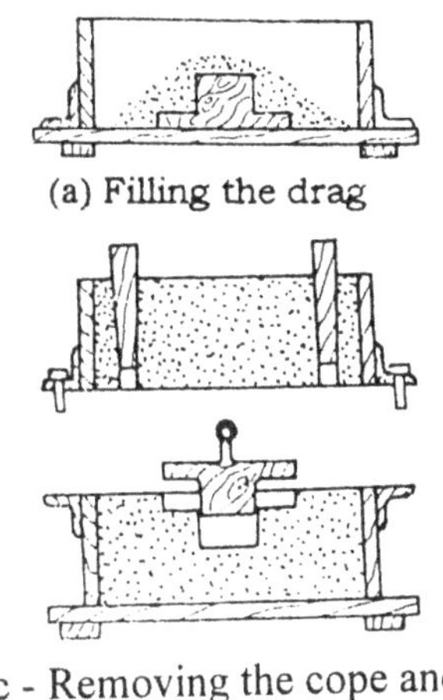

(a) Filling the drag

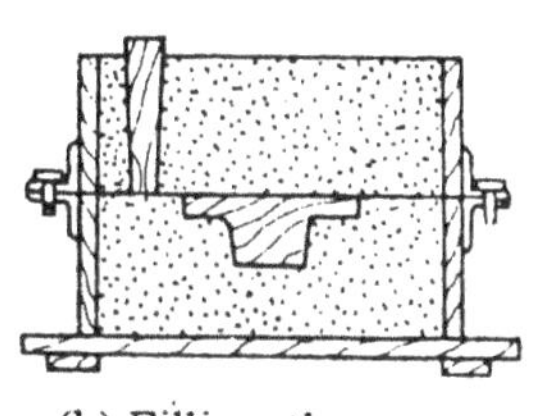

(b) Filling the cope

c - Removing the cope and
drawing the pattern

d - Completed mould

a - Moulding for single piece pattern

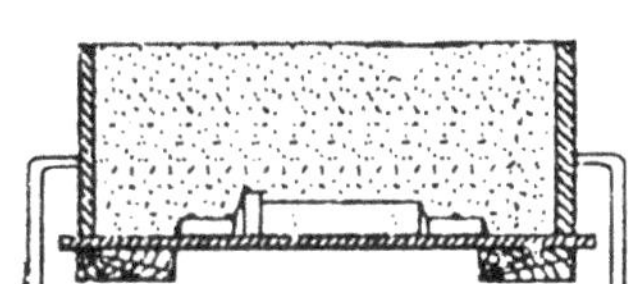

(a) Half pattern rammed-up in the

inverted drag

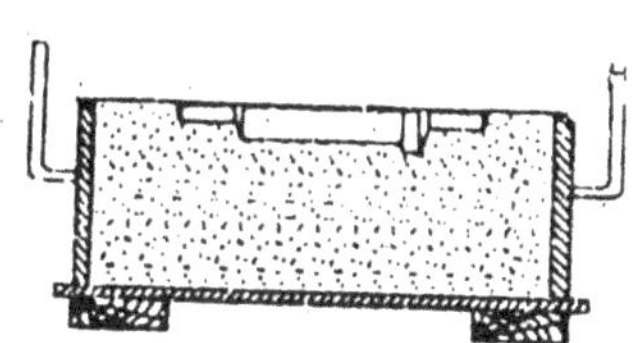

(b) Drag flask turned up

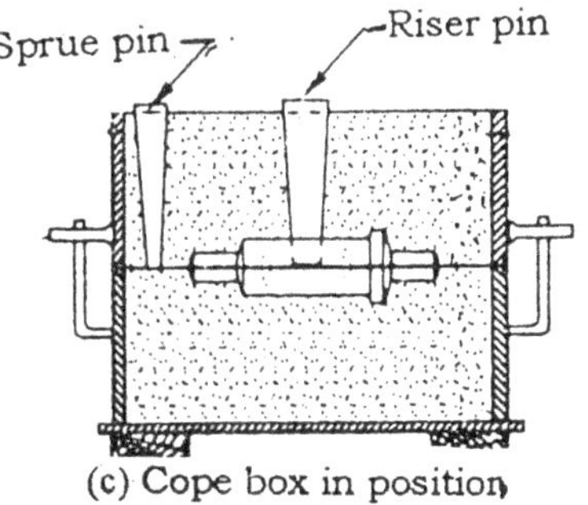

(c) Cope box in position

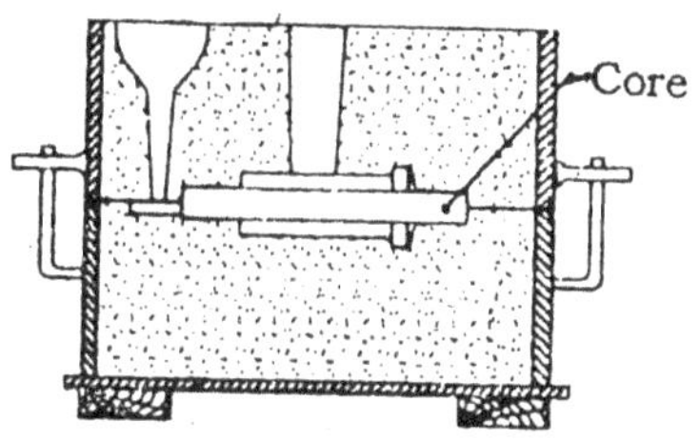

(d) Completed mould
with core

b - Moulding for split pattern

When you are good to others, you are best to you

Recommended Books

a. Workshop technology books

1. Shop theory - Anderson.

2. Elements of workshop technology Vol. I - Hajara Chowdary.

3. Elements of workshop technology Vol. I - Raghuvamshi

b. "Best seller books". On personality development.

1. The Leader in you - Dale Carnegie Associates

2. The Engineer in you - Prof. P.Kannaiah, Prof. K.L. Narayana & Prof. K.Venkata Reddy

3. Read and grow rich - Burk Hedges

4. Grow rich with peace of mind - Nepolen Hill

5. Skill with people - Les Gibblins

6. Especially for you -   Essin

7. Who stole the American dream - Burk Hedges

8. Seven habits of highly effective poeople - Stephen R. Covey.

9. Rich Dad, poor dad - Robert T.Kiyosaki

Exercise - 7.1

Aim : To prepare a mould for a flange.	Ex. No. :
	Date :

Sketch :

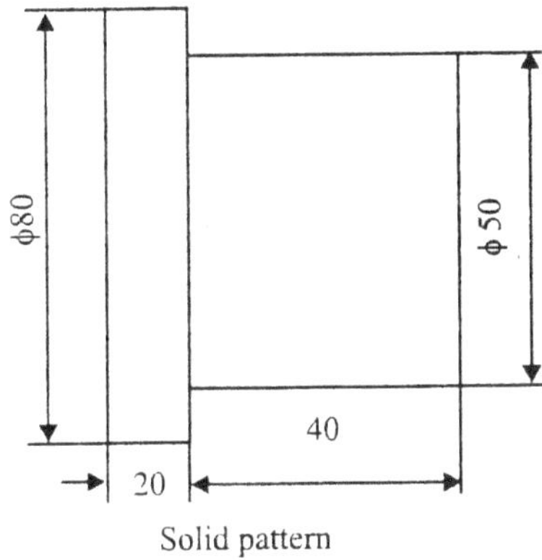

Solid pattern

Material : Moulding sand which contains, sand, clay and water plus additives to give strength.

List of tools used :

Sequence of operations with stages of moulding :

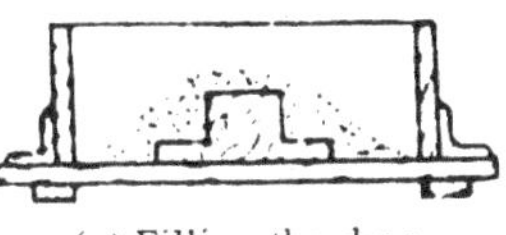

(a) Filling the drag

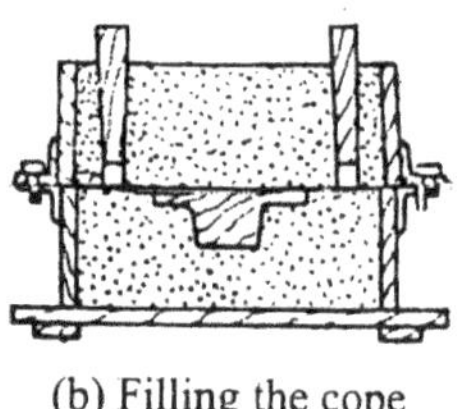

(b) Filling the cope

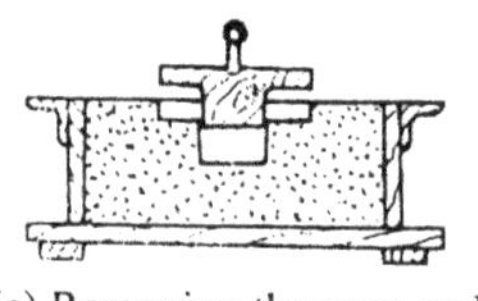

(c) Removing the cope and
drawing the pattern

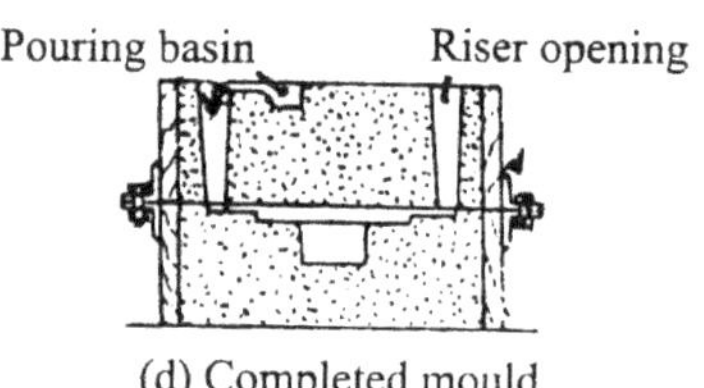

(d) Completed mould

Safety precautions :

Staff signature

Exercise - 7.2

Aim :	Ex. No. :
	Date :

Sketch of pattern :

Material :

List of tools used :

Sequence of operations :

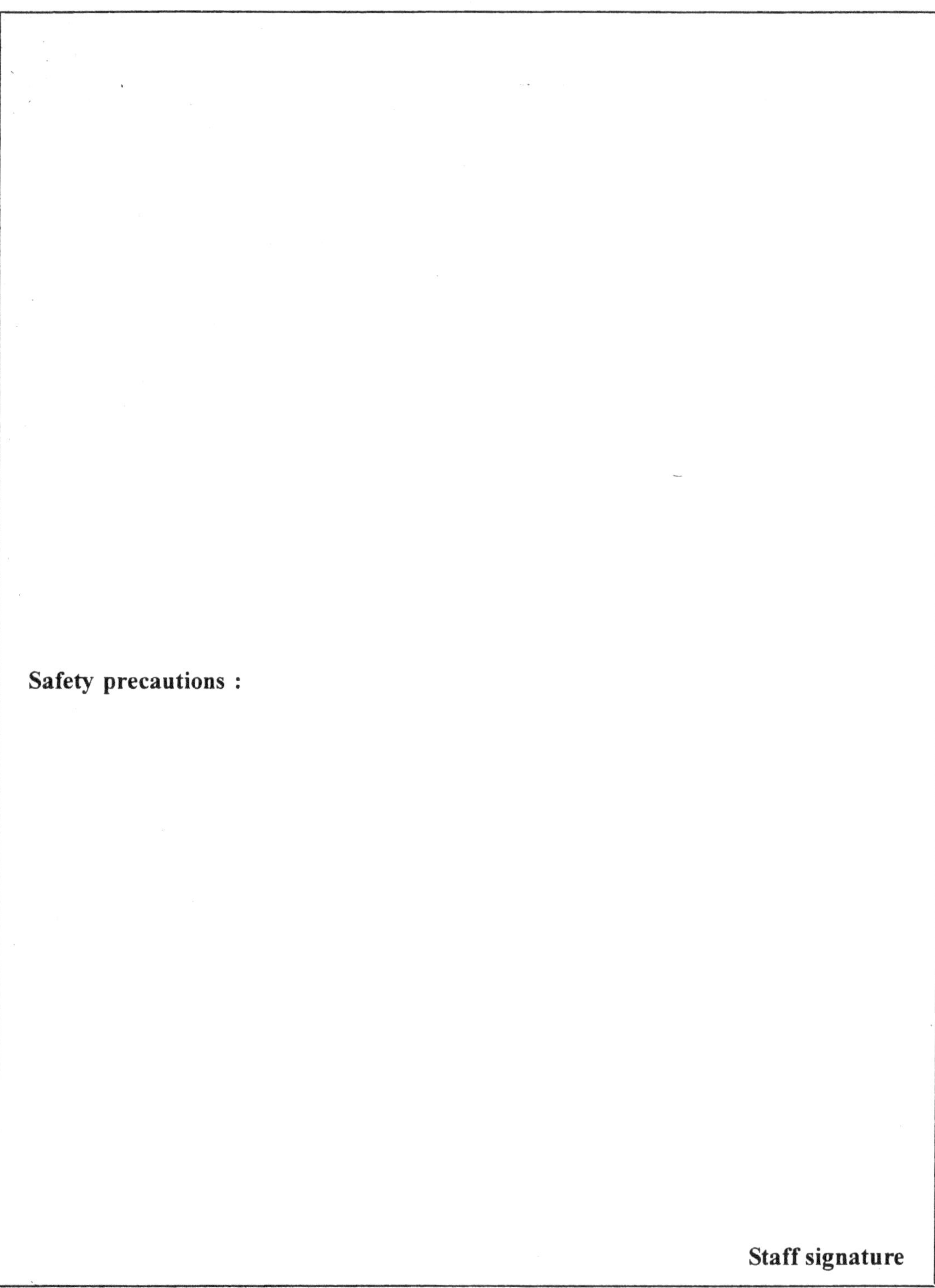

Safety precautions :

Staff signature

Exercise - 7.3

Aim :	Ex. No. :
	Date :

Sketch of pattern :

Material :

List of tools used :

Sequence of operations :

Safety precautions :

Staff signature

Exercise - 7.4

Aim :	Ex. No. :
	Date :

Sketch of pattern :

Material :

List of tools used :

Sequence of operations :

Safety precautions :

Staff signature

CHAPTER - 8

MACHINE SHOP - Lathe only

INTRODUCTION

Manufacture of components involves one or more of the following production processes; (i) Casting processes, (ii) Forming processes, (iii) Fabrication processes and (iv) Material removal processes. In a machine shop material is removed to the required shape on various machine tools by cutting tools.

The most common conventional machine tools found are: Lathes, Shapers, Slotters, Planers, Drilling machines, Milling machines and Grinding machines, where material is removed in chip form by various cutting tools.

LATHE

A lathe is a versatile machine used to cut and shape the material by revolving the work against cutting edge of a cutting tool. The work is clamped either in a chuck or face plate fitted on the lathe spindle or between the centres. The cutting tool is fixed in a tool post, mounted on a movable carriage that is positioned on the lathe bed. The cutting tools can be fed into the work either length wise or crosswise.

Principal parts of a lathe

The principal parts of a lathe are shown in figure

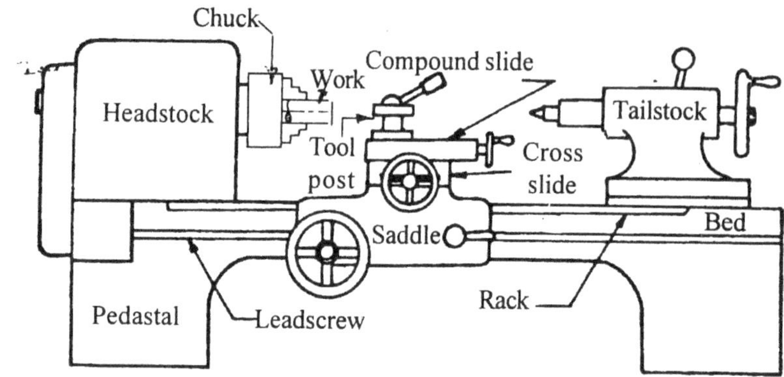

Parts of a lathe

Bed

Bed is an essential part of a lathe, which must be strong, rigid and free from vibrations. It carries all parts of the machine and resists the cutting forces. The carriage and the tailstock move along the guide ways provided on the bed. It is made of grey cast iron because it has the above properties.

Head stock

It contains cone pulleys, V-pulleys or gears to provide the necessary range of spindle speeds. Headstock provides the speed gearbox, which rotates the main spindle at selected revolutions per minute (rpm) depending on the cutting speed (m/mt) recommended for a particular · operation.

Tailstock

It is used to support the right hand end of a long workpiece. It may be clamped in any position along the lathe bed. The tailstock spindle has an internal Morse taper to receive the dead centre that supports the work. Drills, reamers, taps etc. may also be fitted into the tailstock spindle for performing operations such as drilling, reaming and tapping.

Carriage or Saddle

Carriage is used to control the movement of the cutting tool. The carriage assembly consists of the longitudinal slide, cross slide, the compound slide and the apron.

The cross slide moves across the length of the bed and perpendicular to the axis of the spindle. This movement is used for facing and to provide the necessary depth of cut while turning.

The apron, which is bolted to the saddle, is on the front of the lathe and contains the longitudinal and cross slide controls both manual and automatic.

Compound rest

· Compound rest supports the tool post. By swiveling the compound rest on the cross slide, short tapers may be turned to desired angles. Material is cut by moving the compound slide. The length of the taper is limited to the travel of the compound slide.

Tool Post

The tool post holds the tool or the tool holder, which may be adjusted, to any working position. It may be a single tool post that can hold one tool at a time or square tool post with provision to hold four tools. In some lathes quick-change tool posts are also provided.

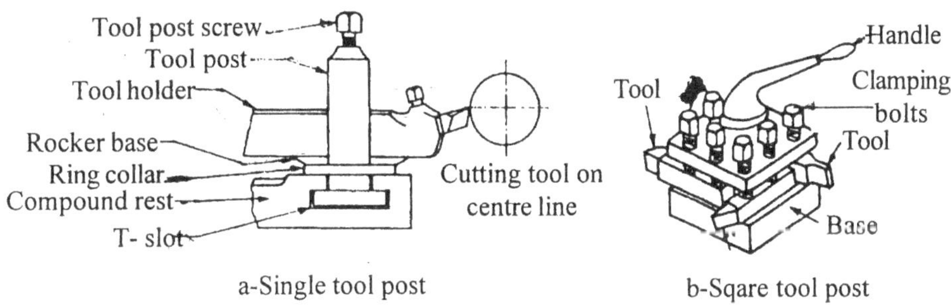

a-Single tool post   b-Sqare tool post

Tool posts

Feed gearbox

It is also called Norton gearbox. It is connected to the lathe spindle through change gears and tumbler gears to provide automatic rotation to the feed rod or lead screw.

Lead screw (feed rod)

It is a long threaded shaft located in front of the carriage, running from the head stock end to the tailstock. It is connected to the spindle through the feed gearbox and change gears for providing movement of the tool either for automatic feeds or for cutting threads. In some lathes feed rod and lead screw are separated. The thread of the lead screw is made of ACME type for easy engagement of half nut.

Centres

There are two types of centres known as dead centre and live centre. The dead centre is fixed in lathe tailstock spindle and the live centre in the headstock spindle. During turning between centres, the dead centre does not revolve with the work, while the live centre revolves with the work

WORK - HOLDING DEVICES

Three jaw chuck

It is a work-holding device having three jaws (self-centering), which will close or open with respect to the chuck centre, centre line or the spindle centre. It is used for holding regular objects like round bars, hexagonal rods etc. A three jaw chuck is provided with a set of external (reverse) jaws also.

Four jaw chuck

In a four jaw (independent) chuck, all jaws have independent movement. It is used for holding square, octagonal or irregular shaped works.

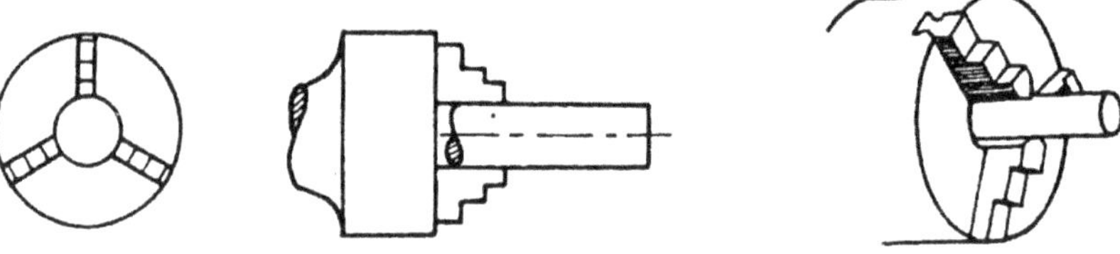

a- Three jaw chuck

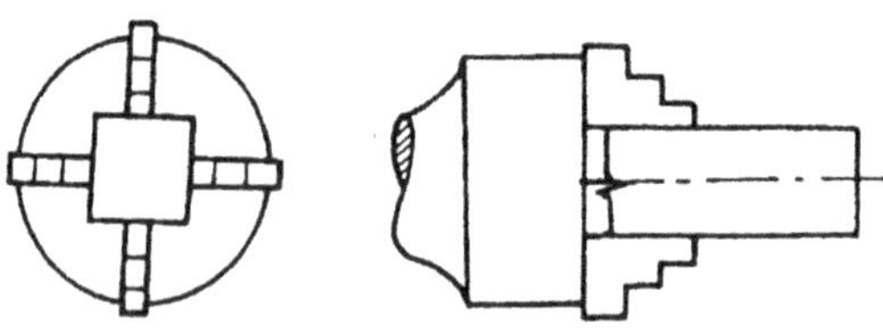

b- Four jaw chuck

Chucks

Face Plate

It is a plate of large diameter used for turning operations. Certain types of works that can not be held in chucks are held on the face plate with the help of various accessories or fixtures.

Lathe dogs and driving plate

These are used to drive a work piece that is held between centres. These are provided with an opening to receive and clamp the work piece and dog tail. The tail of the dog is carried by the pin provided in the driving plate fixed on the spindle nose and rotates along with it.

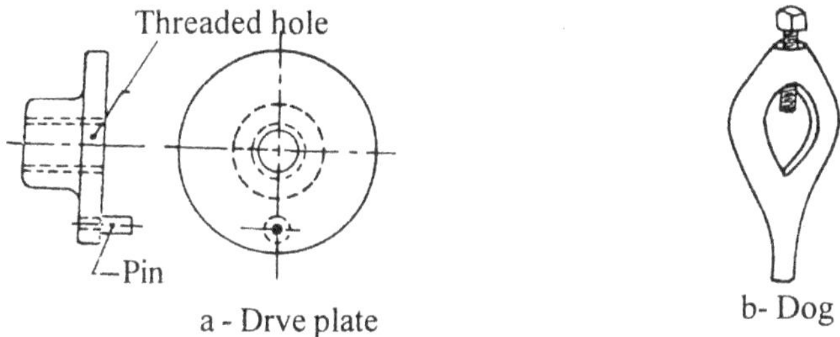

Drive plate and dog

CUTTING PARAMETERS

Cutting speed

Cutting speed is defined as the speed at which the material is removed and it is specified in meters per minute. It depends on the work piece material, tool material, feed, depth of cut, type of operation such as rough turning, finish turning, threading etc. and many other cutting conditions. For example, the cutting speed recommended for turning operation on mild steel with high-speed steel tool is 25m/mt. If, instead of HSS tool, carbide tool is used the cutting speed recommended is about 75 m/mt. On the other hand, instead of mild steel the work material is Aluminum say with the same HSS tool the cutting speed recommended is 150m/mt. The cutting speed recommended for various combinations are given in Table.

Knowing the cutting speed for a particular operation and the diameter of the work, th spindle speed in revolutions per minute (RPM) is calculated from the relations,

$$\text{Spindle rpm} = \frac{\text{Cutting speed x 1000}}{\pi D}$$ Where D is the diameter of the work piece in mm.

Table Average Cutting Speeds in metres per minute for turning with H.S.S and carbide tipped tools

Material to be machined	High Speed steel tools	Cemented Carbide tipped tools
Alluminium and aluminium alloys	120 - 300	300 - 900
Brass	90	220
Cast Iron (grey)	25	90
Cast Iron (malleable)	28	80
Copper	25 - 60	120
Bronze	20	60 - 120
Monel Metal	20	55
Nickel	18 - 20	50
Mild Steel	25	75
Medium Carbon Steel	22 - 24	60 - 120
High Carbon Steel	15	60
Alloy Steels	4 - 9	12 - 45

Problem : Find the rpm of the spindle to rough turn a mild steel job of 30 mm diameter with HSS tool.

Solution : From Table the recommended cutting speed for turning on M.S with HSS tool is 25 m/mt

$$\text{Since } C.S = \frac{\pi DN}{1000}; \quad N = \frac{25 \text{ x } 1000}{\pi \text{ x } 30} = 265 \text{ rpm}$$

Note: The nearest rpm available on the lathe speed gearbox is selected.

Feed

Feed is the distance traversed by the tool during one revolution of the work. It may be along the bed for longitudinal feed or across the bed for cross feed. It is normally specified in mm per

· revolution. It is also specified in meters per minute. The value of feed chosen mostly depends on the depth of cut and finish of the work desired.

Depth of cut

It is the movement of the tip of the cutting tool, from the surface of the work piece and perpendicular to the lathe axis. Its value depends on the nature of operation like rough turning or finish turning, feed rate etc. Depth of cut is given accurately by making use of the graduation Collar.

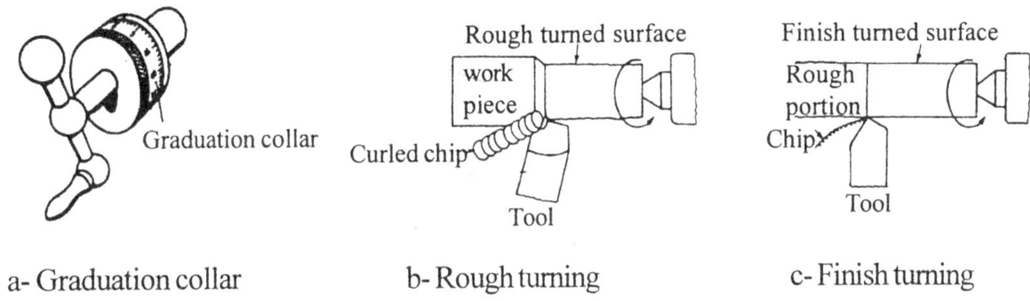

a- Graduation collar	b- Rough turning	c- Finish turning

TOOL MATERIALS

General-purpose hand cutting tools like chisels, knives, axes, etc. are usually made from carbon steel or low alloy tool steel. The single point lathe cutting tools are made of high alloy tool steels know as high-speed steel (HSS). The main alloying elements in 18-4-1 HSS tools are 18 percent Tungsten, 4 percent Chromium and 1 percent Vanadium apart from 0.7 percent carbon. About 5 to 10 percent Cobalt is also added to improve the tool hardness at high temperatures during working.

Cemented carbide tipped tools, fixed in tool holders are mostly used is production shops, which serves at almost three times the cutting speed when compared to HSS. The common commercial brands of cemented carbide tools are T-Max, L & T etc.

TOOL GEOMETRY

A single point cutting tool used on lathe may be considered as a simple wedge. Basic angles of a simple turning tool and lathe tool holders used are shown in figure. Tool angles recommended are given in the Table.

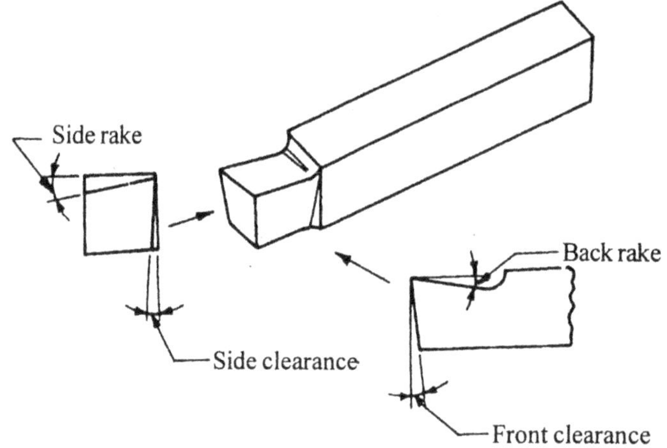

Basic tool angles

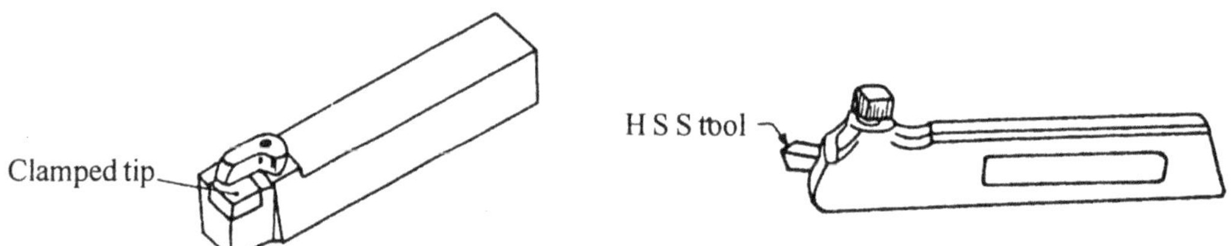

Lathe tool holders

Table: Average values of Tool angles for single point Lathe Tools

Material to be machined	High Speed steel tools			Cemented Carbide tipped tools		
	Top Rake	Side rake	Front clearance	Top rake	Side rake	Front Clearance
Aluminium and Aluminium alloys	20^0-40^0	15^0	8^0	10^0-20^0	15^0	8^0
Brass	0^0	0^0	6^0	0^0	10^0	8^0
Cast Iron (grey)	10^0	12^0	6^0	6^0	10^0	8^0
C I (malleable)	8^0	20^0	6^0	0^0	10^0	6^0
Copper	20^0	30^0	10^0	4^0	15^0	10^0
Carbon steel	10^0 - 12^0	10^0 - 12^0	8^0	0^0	6^0	6^0

A leader owns responsibility (response-ability).

LATHE OPERATIONS

Lathe being a versatile machine, it can do most of the operations done by other machines also with suitable attachments. The common operations that can be done on a lathe are shown in figure. As mentioned earlier, before doing any operation the spindle rpm is selected based on the cutting speed recommended for the operation.

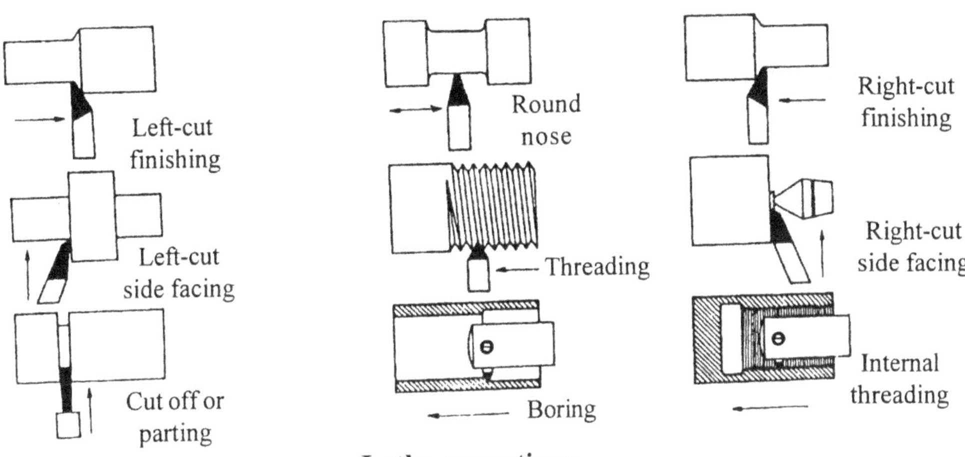

Lathe operations

Taper turning

Short tapers may be cut on the lathe by swiveling the compound rest to the required angle. Here, the cutting tool is fed by means of the compound slide handle. The taper angle is calculated from

the relation,
$$\tan \theta = \frac{D - d}{2L}$$

Where θ is the semi-cone angle, D is big diameter, d is small diameter of taper and L is taper length.

Note : Long tapers may be obtained by means of taper turning attachment or by off- setting the tail stock.

Knurling

Knurling is done to provide rough surface on cylindrical parts for hand grip. This is required on handle knobs, screw knobs of instruments like micrometers etc. Depending on the roughness required smooth, medium and coarse knurling is done on lathe, using knurling tool holders.

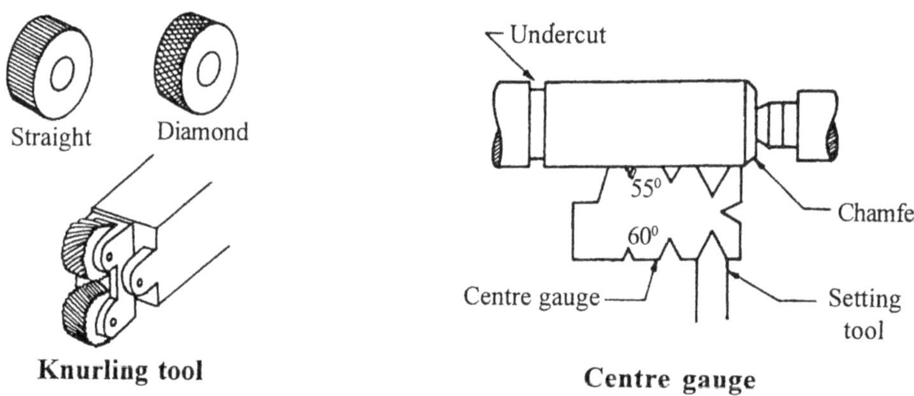

Thread cutting

Threads may be cut either on the external or internal cylindrical surfaces. A specially shaped cutting tool known as a thread cutting tool is used for this purpose.

Thread cutting on a lathe is performed by traversing the cutting tool at a definite relation to the rate at which the work revolves. The saddle which controls the tool movement is in turn traversed by the movement of the lead screw nut along the lead screw. If the driving gear on the spindle and driven gear on the lead screw are of the same size and with a single idle gear in between, the lead screw and the spindle revolve at the same speed, producing threads on the work with the pitch same as that of the lead screw. If the driven gear is double the size of the driver gear, the lead screw revolves at half the spindle speed, producing threads with the pitch, half that of the lead screw.

Centre gauge

Centre gauge is used for setting the thread cutting tool, square to the spindle axis. It is also used for checking the thread cutting tool angles while grinding.

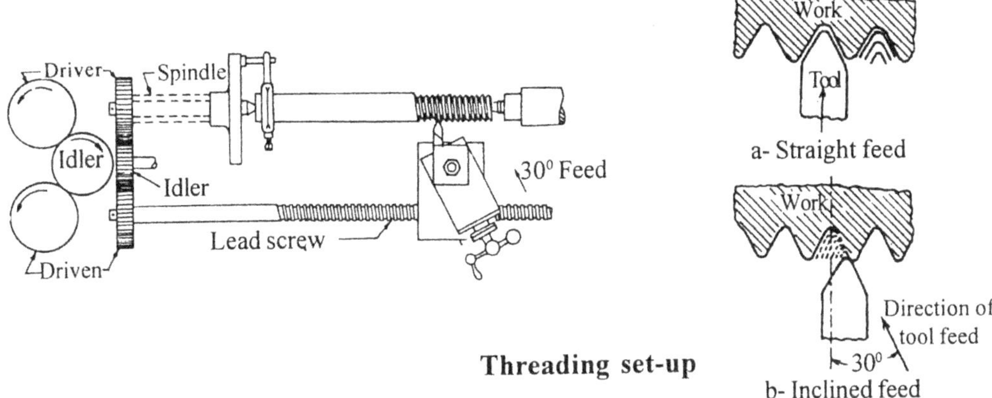

Example : To find the change gears to cut a thread of pitch 2mm on a lathe having a lead screw of 2.5 mm pitch.

$$\frac{\text{Driver}}{\text{Driven}} = \frac{2}{2.5} = \frac{32}{40}$$

Motion is transmitted between the two gears by an intermediate idler gear. The size of the idler has no effect on the gear ratio.

The engagement of half nut with the lead screw makes the carriage and the tool tip follow the path of the helix angle of the thread being cut. When the tool reaches the end point, the half nut is disengaged. The tool is withdrawn from the work by rotating the cross slide by one revolution and the saddle is brought back to starting point. Next cut is given by engaging the half nut after giving the depth of cut either in the compound slide or cross slide. This procedure is repeated number of times until the thread depth is reached.

Note

1. Tool should be ground accurately and set square to the spindle axis using **centre gauge**
2. Machine is run at 1/3 the **cutting speed** for turning, ie. **10m/mt**
3. Engagement of half nut at the correct mark is important for obtaining accurate thread **form, fit** and **finish.**
4. **Pitch gauge** or nut is used to check the form, finish and fitting of the thread.
5. Lubricant is used for better cutting conditions.

Important note

After the tool is brought back to the starting point, setting it to the previous cut position is very important. This is done by making a chalk piece mark on the cross slide graduation collar after each cut. If by chance the chalk mark on the graduation collar is crossed, it will result in excess cut and bending of job etc. Normally when the chalk mark crosses there is a tendency of the student to rotate the cross slide wheel in the reverse direction to the extent crossed to correct it and proceed with the next cut.

Actually, by this correction the chalk mark appears to have come back to the correct position but the tool remains to be in the advanced position. This is due to the backlash between the cross slide lead screw and the nut due to wear. Therefore, whenever such a mistake happens, it is advisable to rotate the cross slide wheel back again by one revolution and bring it forward again by one revolution, taking care to coincide the chalk marks exactly this time. Thus eliminating backlash.

Combination centre drill (C.C.Drill)

C.C Drill

Combination centre drill is used to make centre holes on the ends of the job turned for supporting on the dead centre. Centre hole is also made before drilling holes on lathe.

Note : CC drill size is selected based on the diameter of the job. It is important that sufficient taper length is drilled, so that the job rests on the dead centre uniformly.

Example 2 : To find the change gears to cut a B.S.W thread of 10 TPI (threads per inch) on a lathe having a lead screw with 8 TPI.

Solution :

$$\frac{\text{No of teeth on the driver}}{\text{No. of teeth on the screw}} = \frac{\text{Pitch of the job}}{\text{Pitch of the L.S}}$$

$$\frac{8}{10} = \frac{32 \text{ teeth}}{40 \text{ teeth}}$$

Example 3 : To find the change gears to cut a metric thread of 2mm pitch on a lathe having a lead screw with 8 TPI.

Solution :

$$\frac{\text{Driver}}{\text{Driver}} = \frac{\text{Pitch of the job}}{\text{Pitch of the L.S}}$$

$$\frac{2}{1/8 \times 25.4} = \frac{2 \times 8}{25.4}$$

If we take care of the beginning, the end will take care of itself.

$$\frac{\mathbf{48}}{24} \times \frac{40}{127}$$ Compound gear

The arrangement is shown in Figure.

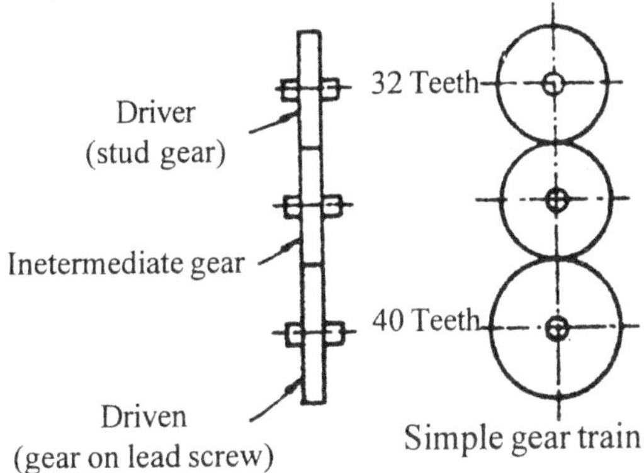

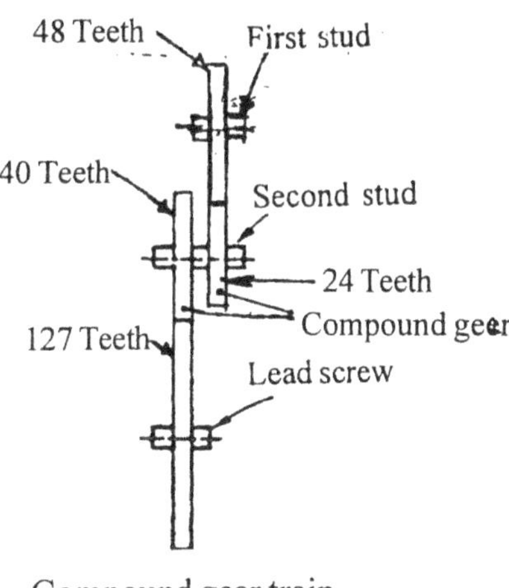

Note

(i)   A thread is said to be right hand when the bolt advances with the rotation of the nut in the clock wise direction from top; and left hand when it is anti-clockwise.

(ii)   The movement of the carriage is from right to left (towards the chuck) for right hand thread cutting or turning and from left to right for left hand cutting.

(iii) For left hand turning or threading the position of the tumbler gears is changed to change the direction of rotation of the feed rod or lead screw as the case me be or job the chuck. rotating in the same direction (ie) anti-clockwise.

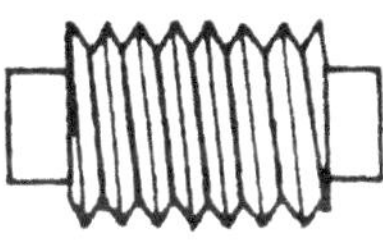

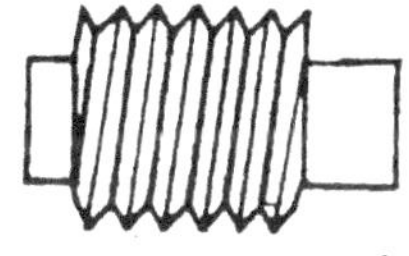

Thread pitch gauge   Right hand thread   Left hand thread

Shop floor inspection of the thread is done with the help of a thread pitch gauge. A V-thread is specified by its nominal diameter and pitch apart from its standard form say BSW, Metric, etc. (See Fitting shop).

Example (i)   ¾" BSW - Means a British Standard Witworth of ¾" outside diameter and 10 threads per inch or 1/10 of an inch pitch. The thread has 55° V-shape

(ii)   M 20 x 2.5 - Means a Metric thread as per BIS (ISO also) with nominal diameter of 20mm and pitch of 2.5 mm. The thread has 60° V-Shape. Separate pitch gauges are available for BSW and Metric threads with markings on the leaves.

Gentleness and friendliness are stronger than force and fury.

Model Record Sheet

Aim To turn a model as per the Figure below

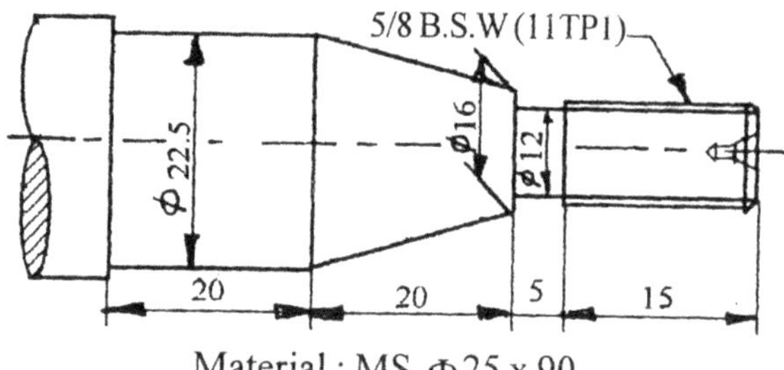

Material : MS ϕ25 x 90

Tools required

Round nose tool, parting tool, threading tool, steel rule, vernier calipers, nut.

Sequence of operations

(1)   Hold the job in the chuck leaving 30mm outside the chuck jaws.

(2)   Set the round nose tool in the tool post to the height of centres of the lathe or tailstock dead centre point.

(3)   Set the spindle to about 300 rpm.

(4)   Face the end.

(5)   Turn ϕ 22.5 x 60mm length in stages.

(6)   Put the centre drill.

(7)   Support the job on the tail stock dead centre.

(8)   Turn the remaing length to 23.5mm

(9)   Calculate the taper angle.

(10)   Set the compound rest to the angle calculated and turn the taper in steps giving feed in the compound slide.

(11)   Fix the parting tool in the tool post.

(12)   Turn the undercut to ϕ12 x 5 mm.

(13) Fix the threading tool.

(14) Calculate the change gears and fix them. (Stud gear 32 teeth and screw gear 44 teeth).

(15) Engage the backgear.

(16) Engage the tumbler gear to give feed to the lead screw.

(17) Change the cutting speed to **100 rpm**.

(18) Cut the thread in steps.

(19) Check the **Form, Finish** and **Fit** using a nut or pitch gauge.

Note Write the sequence of operations in past tense in your records as done by you.

Safe Practices

(1) Dress properly. Roll up full sleeves. Don't wear Jewellry (chains) or neck tie: Avoid long hair too.

(2) Clamp the work firmly and use correct speed, feed and depth of cut.

(3) Stop the machine before making measurement or adjustments.

(4) Never change the speeds while the machine is running

(5) Do not stop the machine by reversing the direction of rotation.

(6) Use a brush and not hands to clean the chips from work or swarf on tool chip.

(7) Never clean moving machines and never apply a rag or cotton waste to revolving work.

(8) Keep proper posture and avoid watching too close to the cutting place. The chips may fly and damage your eyes.

(9) Never leave the chuck key in the chuck.

(10) Call on doctor for removal of chips or swarf that had entered the eye. Never rub your eye as it may damage the eye.

OBJECTIVE QUESTIONS

Fill in the blanks or choose the correct answer

(1)   While turning, the chuck rotates in the clockwise / anti-clockwise direction, when viewed from the tailstock side.

(2)   The main alloying elements in 18-4-1- HSS tool are _____ _____ _____ .

(3)   The carbon content in HSS tool is _____ .

(4)   The carbon content in mild steel is _____ .

(5)   The carbon content in high carbon steel is _____ .

(6)   For doing taper turning, feed is given by moving compound rest / compound slide.

(7)   An octagonal bar can / can not be held in a three jaw chuck.

(8)   The rpm of the lathe spindle depends on the _____ speed of operation

(9)   When feed rate increases, the finish on the work increases / decreases.

(10)   Slender jobs are supported on _____ while turning.

(11)   _____ is used to support the lengthy jobs.

(12)   _____ centre is used in head stock.

(13)   The taper on the dead centre is called _____ taper.

(14)   A thread is defined by its _____ and _____ .

(15)   _____ tool is used for cut-off.

(16)   To obtain slow spindle speeds _____ gears are used.

(17)   Internal taper is done by a _____ tool.

(18)   Lead screw has ACME/square threads.

(19)   Cross slide is operated to do _____ operation.

(20)   The angle of the metric thread is _____ degrees.

(21)   To set the threading tool accurately _____ is used.

A person who gets elected is not a leader but a follower of those electing him.

CHAPTER - 9

PLUMBING

INTRODUCTION

Pipes are used for carrying fluids such as water, steam, gas, oil, etc., from one place to another. As pipes are made in standard lengths, the desired length of a pipe may be obtained by joining them. The type of joint used depends upon the material of the pipe and the purpose for which it is used.

Generally, pipes are made of cast iron, wrought iron, steel, brass or copper. The material selection is based on the nature of the fluid to be conveyed, viz., pressure, temperature, chemical properties, etc. Now-a-days PVC pipes are extensively used with ease for various purposes.

The terms pipe and tube are synonymously used both in specifications and application as well. The standard codes of practice followed for specification of pipes is British system only. However, the sizes may be converted into meteric units by taking 1 inch = 25.4 mm as a multiplying factor. there does not exist a separate metric system of specifiction.

The pipes are classified based on nominal size and wall thickness as standard (STD), extra strong (XS) and double extra strong (XXS). The external diameter of the pipe in all the three cases is same, however, the wall thickness varies, depending on the fluid pressure in the pipe. Hence, the internal diameter becomes smaller if the pipe is XS or XXS.

There exists a relation between the external diameter of the pipe (actual) and the nominal size fo the pipe, as spesfied in the standards. The standard pipes are produced of nominal sizes 1/8 inch to 36 inches actual size. However, upto 12 inches nominal size, the outside diameter (actual) is greater from the nominal size specification. From 14 inches and above, the nominal size and outside diameters (actual) are same i.e., pipe of nominal size 14 inches, has outside diameter 14 inches only.

The true art of memory is the art of attention.

The pitches given in TPI for BSW pipe threads are standard as given below:

Pipe size, mm (standard in inches)	TPI
20 (1/2″)	14
25 (3/4″)	14
32 to 63 (1″ to 2″)	11 1/2
75 to 63 (2 1/2″ to 6″)	8

G.I pipe fittings

Wrought iron is coated with zinc by a process known as galvanising to make galvanised iron (G.I) pipe fittings which do not rust readily like C.I fittings. Some of the commonly used G.I pipe fittings are shown figure below.

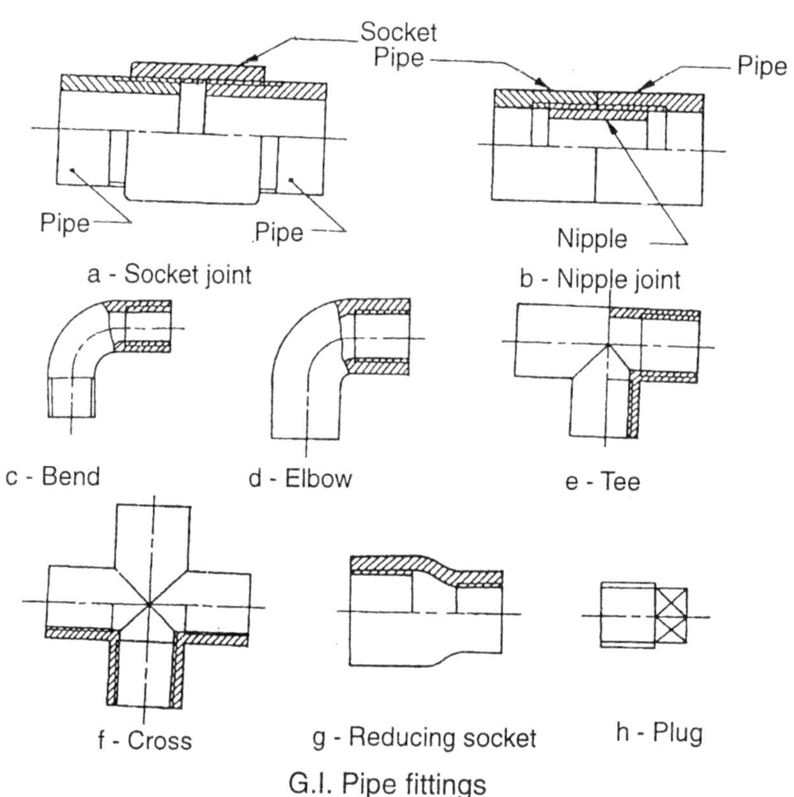

G.I. Pipe fittings

A mans real possession is his memory.

1. **Coupler :** For making-up the length, in genral, wrought iron and steel pipes are joined by means of a socket or coupler. It is a small pipe with internal threads throughout, used to connect the pipes having external threads at their ends .

2. **Nipple :** A nipple is a small pipe, threaded throughout on the outside. For making-up the length, the nipple is screwed inside the internally threaded ends of other pipes or pipe fittings. This type of joint, causes restriction to the fluid passage.

3. **Bends, elbows, tees and crosses :** The fittings are used either to connect or branch off the pipes at right angles.

4. **Reducing socket :** It is used to connect two pipes of different diameters.

5. **Plug :** It is lused to close the end of a pipe with internal threads. For the same purpose, a plug with internal threads can also be used to close a pipe end with external threads.

PVC pipes and fittings

Plastic pipes and pipe fittings made of poly Vinyl chloride (PVC) are extensively used for agricultural, industrial and domestic use. These pipes are made in sizes varying from 20mm to 315mm diameter as per IS 4985:81.

Salient features

PVC pipes exhibit the following salient features:

(1) Smoother bore in comparision to C.I, G.I and also cement pipes, thereby better flow characteristics.
(2) Seemless, strong and resilient.
(3) Light weight, offering total economy in handling, transportation and installation.
(4) Resistence to chemical, electrolytic and galvanic corrosion.
(5) Odourless and hygenic for transporting potable water, as they do not subject to contamination.
(6) Maintenance free.
(7) Long-lasting-PVC is free from weakening caused by scale formation rusting, weathering and chemical action and hence more durable for rated working conditions.

Properties

The following properties are expected form pipes made as per ISI specifications:

1. Thermal conductivity - 4×10^{-4} cal/hrCm cm^2/^{0}C/Cm
2. Co-efficient of linear expansion - 5.0 to 6.0×10^{-6} mm/^{0}C
3. Specific gravity - 1.41 gms/cm^3
4. Combined flexural and compressive strength - 600 kg/cm^2
5. Impact strength at 20^{0}C-3kg/cm^2
6. Electrical resistance - 10^{14} ohms cm
7. Modulus of elasticity - approximately 30,000 kg/cm^2
8. Vicat softening point - 81^{0}C

Dimensions of PVC pipes as per IS 4985-1981 are given in Table It may be noted that the pipe size indicates the outside diameter of the pipe.

Table: Dimensions of PVC pipes
All dimensions in millimetres

		Wall thickness for Working Pressures							
Outside diameter	Tolerance on outside diameter	Class 1 2.5kgf/cm²		Class 2 4kgf/cm²		Class 3 6kgf/cm²		Class 4 10kgf/cm²	
		Min.	Max.	Min.	Max.	Min.	Max.	Min.	Max.
20	+0.3	-	-	-	-	-	-	1.1	1.5
25	+0.3	-	-	-	-	-	-	1.4	1.8
32	+0.3	-	-	-	-	-	-	1.8	2.2
40	+0.3	-	-	-	-	1.4	1.8	2.2	2.7
50	+0.3	-	-	-	-	1.7	2.1	2.8	3.3
63	+0.3	-	-	1.5	1.9	2.2	2.7	3.5	4.1
75	+0.3	-	-	1.8	2.2	2.6	3.1	4.2	4.9
90	+0.3	1.3	1.7	2.1	2.6	3.1	3.7	5.0	5.7
110	+0.4	1.6	2.0	2.5	3.0	3.7	4.4	601	7.0
140	+0.5	2.0	2.4	3.1	3.8	4.8	5.5	707	8.7
160	+0.5	2.3	2.8	3.7	4.3	5.4	6.2	808	9.9
180	+0.6	2.6	3.1	4.2	4.9	6.1	7.0	909	11.1
200	+0.6	2.9	3.4	4.6	5.3	6.8	7.7	11.0	12.3
225	+0.7	3.3	3.9	5.2	6.0	7.6	8.6	12.4	13.9
250	+0.8	3.6	4.2	5.7	6.5	8.5	9.6	13.8	15.4
280	+0.9	4.1	4.8	6.4	7.3	9.5	10.7	15.4	17.2
315	+1.0	4.6	5.3	7.2	8.2	10.7	12.0	17.3	19.3

Applications

The following are the typical applications of PVC pipes:

1. Agricultural and lift irrigation schemes
2. Rural and urban drinking water supply schemes
3. Industrial/chemical effluent disposal systems

4. Acids and slurries transportation
5. Telecommunication
6. Bio-gas (Gober-gas)/Natural gas, and oil distribution lines
7. Tube well casings
8. Under-ground or open pipe line/drainage
9. Domestic plumbing and drainage systems
10. Sewage and drainage schemes
11. Air conditioning/industrial ducting
12. Main line in sprinkler/Drip irrigation system

Joining

Threaded PVC pipe fittings should not be over-tightened, as the threads may get damaged. PVC pipes should never be threaded, they are joined by a solvent cement or by suitable threaded fittings. The steps followed for joining the PVC pipes are:

1. The pipes are cut as square as possible. The pipe and socket should be clea   and dry. The surface is cleaned with emery paper before joining.
2. Thick coat of solvent cement is applied on the outer surface of the pipe and also on the inner surface of the socket.
3. The pipe is inserted into the socket and turned through 90°for even distribution of cement.
4. The joint is held firmly without slipping for 2 minutes and allowed to dry.
5. After 24 hours, the pipe lines are ready for use.

PVC fittings

These are available in various sizes from 20 to 160mm made as per IS 7834-75. The description of the fittings and their technical data are given in Table. It may be noted that the fitting sizes indicate the inside diameter of the opening. The dimensions of commonly used PVC pipe fittings are shown in figure, taking 25mm as an example.

The leader is always a part of the answer.

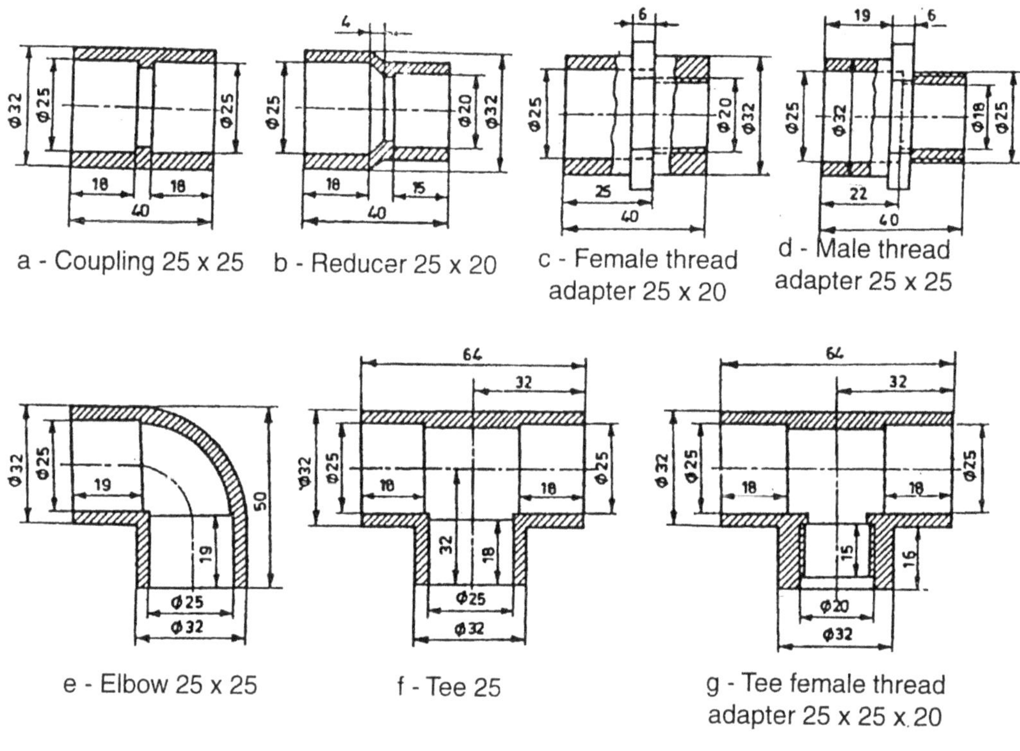

a - Coupling 25 x 25   b - Reducer 25 x 20   c - Female thread adapter 25 x 20   d - Male thread adapter 25 x 25

e - Elbow 25 x 25   f - Tee 25   g - Tee female thread adapter 25 x 25 x 20

P V C Pipe fittings

Tap valve fittings

These are small valves with their outlet open to the atmosphere, whereas the special valves are intermediate fittings between the inlet and outlet to control the flow.

Different types of common taps are shown in Figure. They are made of malleable iron or brass by casting and electroplated for better appearance. Low cost plastic taps are also available in the market.

Table: PVC fittings-technical data

Description	Size in mm. (ID)	Inch equivalent	Application
1. Coupler	20 25 32 40 50 63 75 90 110 160	0.50 0.75 1.00 1.25 1.50 2.00 2.50 3.00 4.00 6.00	These are used for joining two PVC pipes.
2. 90^0 Elbow	20 to 110	0.5 to 4.0	These are used for joining two PVC pipes.
3. 90^0 Bend	20 to315	0.5 to 12.00	These are used where long turns of angle are required. They do not cause anyfriction losses.
4. Equal Tee	20 to160	0.5 to 6.0	These are used for by-pass and taking big service line out of main line.
5.Reducer Tee	20 to160	0.5 to 6.00	These are used for reduced taping from main line.
6. Male threaded adapter (MTA)	20 to140	0.5 to 5.00	These are used to connect the G.I. pipe line/metal line and all types of valves, taps, etc., through a male portion of PVC threaded adaptor.

Every adversity has within it the seed of an equivalent or a greater benefit.

7. Female threaded adapter (FTA)	20 to90	0.5 to 3.00	These are used to connect a PVC pipe line directly to a metal pipe.
8. Reducing FTA	25x20 to 90x75	0.75x0.5 to 3.00x2.50	These are used to connect a PVC pipe line directly to a metal pipe of lower diameter or vice versa.
9. Reducer	2.5x20 to160x140	0.75x0.5 to 6.00x5.00	These are used to convert the service line, into a small line.

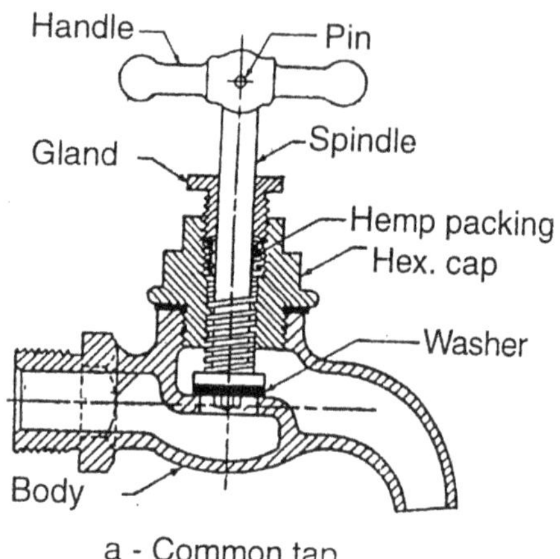

a - Common tap

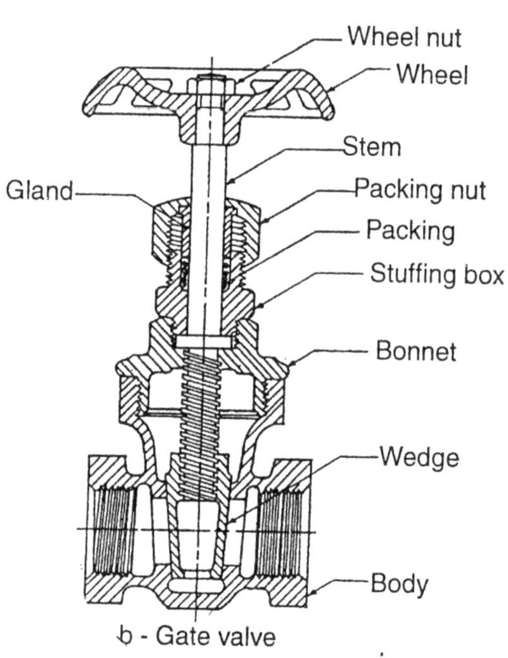

b - Gate valve

IT WORK SHOP

INTRODUCTION

The Computer is a fascinating machine. It is a gate way to a wonderful world of information and myriad applications for the good of human beings. Be it business, academics, defense strategies, budgeting, research, engineering, medicine and space exploration, computers have become established as an indispensable.

The term Computer is derived from the Latin words **"Com"** and **"Putare"**, com means intensive and Putare means calculation. A computer is an electronic equipment which is capable of creating solutions by performing complex processing of information without normal intervention and that too at a tremendous speed.

ORIGIN OF COMPUTERS

About Charles Babbage was known as father of modern computer. For the data storage he used a total number of 70 Vacuum Tubes. To increase the speed of the storage he invented another machine in the year 1850 and it is known as "Analytical Engine". In this machine he used 7000 vacuum tubes. To increase the speed further he used "Character Readers". In this a total of 50,000 long movable parts were used. It was of 2.2 meters height, 15 meters width and of 0.25 km long. Air conditioner was kept at half kilometer distance from the computer.

Definition of Computer

It is a digital electronic storage device which is used to accept the data, process the data and give the output.

(**Data**: It is set of information.)

History of Computers

Invention	Year	Name	Country	Features
I	1501	Abacus	China	Storing Data
II	1617	John Naiper	Germany	Storing Data + Multiply
III	1640	Blasé Pascal	French	Storing Data + Multiply, Add and Subtract
IV	1670	Gott Fred	U.S.A.	Storing Data + Multiply, Add, Sub & Division
V	1820	Charles Babbage	England	Storing Data + Multiply, Add, Sub, Div, Less and Greater

CAPABILITIES OF COMPUTERS

Automation   :   Takes series of information and performs all operations simultaneously.

Speed   :   Can carryout an instruction in a fraction of a second.

Accuracy   :   Produces accurate results provided, data fed is correct and instructions are proper.

Repetitiveness :   Can repeat any operations any number of times without any fatigue.

Storage   :   Stores any amount of information in a very limited space.

Versatility   :   Can perform several independent functions simultaneously at any time in any environment in a desirable fashion.

LIMITATIONS

i.   Lack of commonsense: Like human beings computers do will not have common sense i.e., It cannot perform or take decision on its own, which makes it unfit in some areas.

ii.   Inability to correct: If any instruction is given wrongly the computer cannot correct on its own, it completely depends on instructor.

TYPES OF COMPUTERS

1. Digital Computers   : Used for information processing

2. Analog Computers   : Used for monitoring Temperature, Voltage, Current readings etc.

3. Hybrid Computers   : Combination of Digital and Analog computers is Hybrid computer

GENERATION OF COMPUTERS

First Generation Computers (1942-1955): Voluminous computers. These computers used Vacuum tubes. These computers consume large Power. Also they are less reliable.

Second Generation Computers (1955-1964): Nothing but Transistors. They consume less power and less expensive.

Third Generation (1964-1975): Development over to the transistors, used integrated circuits in place of transistors. They work at higher Speed, Store large amount of data comparatively and less expensive.

Fourth Generation (1975-1989): Large scale integrated chips are used. These computers are used for information processing.

Fifth Generation (1989-till date): These computers are used for parallel processing, possess artificial intelligence. These computers are able to think as human beings.

PARTS OF COMPUTER

(i) Central Processing Unit (CPU)

(ii) Memory Unit

    (a) Primary storage

    (b) Secondary storage

(iii) Input/Output (I/O) Devices:- Keyboard, Mouse, Joystick, Optical pen etc.

Central Processing Unit (CPU)

It is the "brain" of the computer. The Central Processing Unit (CPU) of computer is usually located in Tower-shaped or Horizontal cabinet. The CPU consists of different small components like motherboard, hard disk, RAM, floppy disk drive, sound card etc. It is the most important component of a computer because it can manipulate and change information in memory. It consists of Microprocessor chip which undertakes all the thinking for the PC and runs the programs (series of instructions) according to the user's command.

Arithmetic & Logic Unit(ALU)

Modifies data in response to program instructions. Calculations are performed and all the comparisons are made in the ALU. Such as add, subtract, multiply, compare, etc

Control Unit (CU)

It is like a 'MANAGER' which monitors the entire operations like controlling the flow of data and instructions held within the CPU. Fetches and interprets instructions from the memory and also directs the execution of program instructions

Memory Unit (MU)

Stores the information that the computer can manipulate and stores the programs that manipulate the information. Computer can manipulate both DATA and PROGRAMS

It is the internal storage within the computer itself. Information held in memory can be located in a short time. It is usually only temporary storage.

Memory Organization

All data in the computer are stored as binary numbers (composed of 1's and 0's). It is array of storage locations. Each location designated by a number (address) and contains a number (data/content). The internal storage within computer is divided into two units.

(a)Primary storage: The computer can not be started with out the primary storage.

Examples:

RAM	: Random Access Memory
ROM	: Read Only Memory
PROM	: Programmable ROM
EPROM	: Erasable PROM
EEPROM	: Electrical Erasable Programmable ROM

(b)Secondary storage: It is a permanent storage.

Examples:

Hard Disk

Floppy

PROM	: Programmable ROM
EPROM	: Erasable PROM
EEPROM	: Electrical Erasable Programmable ROM

(b)Secondary storage: It is a permanent storage.

Examples:

Hard Disk

Floppy

Compact Disk

ZIP

DVD

PEN DRIVE

Memory described by

The number of locations (addresses) available

The number of binary digits (bits) stored at each location

Usually 8 digits (8 bits = 1 byte) at each address

1 Kilo = 1024 (210) locations ≈ 1000

1 Mega = 1,048,576 (220) locations ≈ 1,000,000

1 Giga = 1,073,741,824 (230) locations ≈ 1,000,000,000

64MB = 67,108,864 bytes = 536,870,912 bits

Components of Computer

Basic Components	**Optional**
Processor	Printer
Hard disk	Scanner
RAM	Modem
Display/Video card	DVD drive
Keyboard	Speakers

Drives: Hard Disk, Floppy   ZIP drive

Disk and CD-ROM

Mouse

Monitor

PC Case or Cabinet

The PC case is a thin sheet metal enclosure that houses the motherboard, power supply and various drives (HDD, FDD, CD, DVD).

- Cases are offered in two styles, desktop and tower. Today the tower type is predominant. It stands upright and is much taller than it is wide. It is usually placed on the floor next to, or under a desk.

- Tower cases are offered in two basic sizes, one that can fit ATX (12" wide) motherboards and one that can accommodate ATX mini (8.5" wide) motherboards. The number of drive bays offered also varies depending on manufacturer.

- The motherboard and power supply are fixed to the base at the rear of the case. The drives are mounted (hard, floppy and CD/DVD) in enclosures called drive bays at the front of the case. Cases depend upon size (ATX or ATX mini), number of drive bays and the wattage of the power supply.

Motherboard

The motherboard is the main circuit board in a PC. It contains all the circuits and components that run the PC. It is the central or primary circuit board making up a complex electronic system, such as a modern computer. It is also known as a mainboard, baseboard, system board, or a logic board in Apple computers and is sometimes abbreviated as mobo. The basic purpose of the motherboard, is to provide the electrical and logical connections by which the other components of the system communicate.

Major Components in the Motherboard

CPU - the Central Processing Unit is often an Intel Pentium or Celeron or AMD Authelon processor. It is the heart of every PC. All scheduling, computation and control takes place here.

- Chip Set - these are large chip(s) that integrate many functions that used to be found in separate smaller chips on the motherboard. They save space and cost.

- The functions performed by these chip sets are often broken into two devices with one providing an interface from the CPU to the memory and the other providing controllers for IDE, ISA, PCI and USB devices.

A typical desktop computer is built with the microprocessor, main memory, and other essential components on the motherboard. Other components such as external storage, controllers for video display and sound, and peripheral devices are typically attached to the motherboard via edge connectors and cables, although in modern computers it is increasingly common to integrate these "peripherals" into the motherboard. Which is shown below.

Micro processor

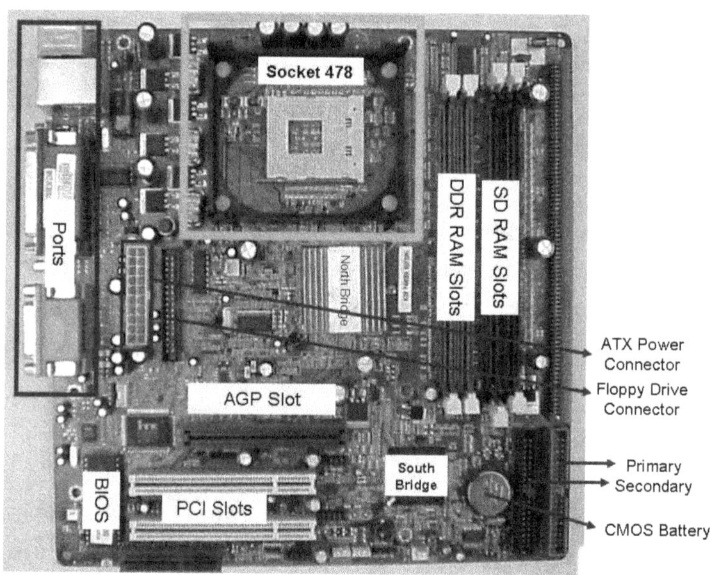

Mother Board

Hard Disk Drive

The HDD installs in one of the 3-1/2 inch internal drive bays in the PC. It is secured by machine screws. It is powered by a 4 conductor cable coming from the power supply. Data to and from the motherboard is carried on a 40-pin IDE (Integrated Drive Electronics) cable.

- Data is stored magnetically on multiple rigid disks that are stacked up on over the other. Small arms with magnetic pickups move rapidly back and forth across the top and bottom surface of each disk in the drive. The sensors float just a few microns above the rotating disk surface and can read and write data at very high rates.

- Most commercially available hard drives rotate at 5400 or 7200 RPM (revolutions per minute) which translates to 90 or 120 revolutions per second respectively. The data transfer rate from the drive to the motherboard is 33 Mbytes/second in bursts. Newer drives are capable of higher speeds up to 66 Mbytes/sec. To use this faster drive, the *PC must have an ATA/66 interface* that is capable of keeping up with it.

Types of HDD

    (i) IDE (Integrated Drive device Electronics)

    (ii) SATA (Serial Data Transfer device)

SATA is fastest storage device than the IDE type. Capacities of range 40 , 80, 120,160,250,300 GB are available in the market.

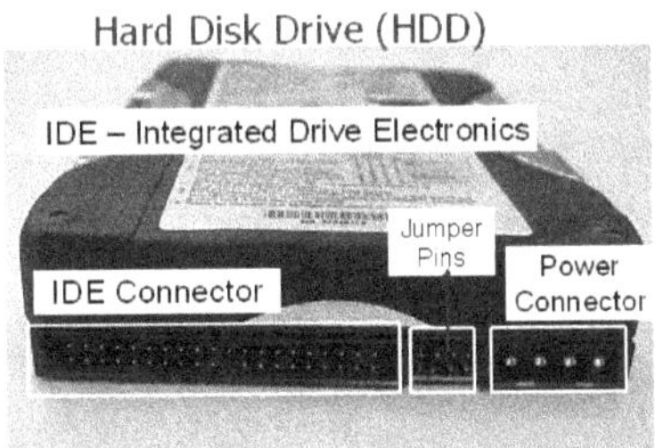

Floppy Disk Drive

- The FDD is installed in one of the *external drive bays* at the front of the PC case and is secured by machine screws. By external means one can access the drive from the outside.

- It is powered by a cable with a 4-pin connector that comes from the power supply.

- It transfers data to and from the motherboard by means of a 34 pin ribbon cable.

- It stores data magnetically on a removable floppy disk. A pickup arm in the drive floats above the disk surface. The arm moves rapidly back and forth across the disk surface as a small magnetic sensor at the end of the arm, reads and writes data on the rotating disk surface.

- Floppy disks hold 1.44 Mbytes, which at one time was a large amount of data. Today many programs and files are much larger than this. In spite of being surpassed in size by CD and DVD, floppy drives are still found on many newer PCs.

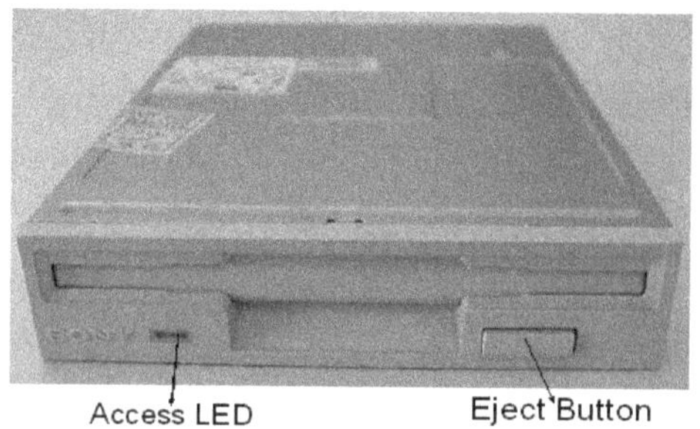

Floppy Disk Drive

Monitor

The monitor of a computer is like a television screen. It displays text character and graphics in colour or in shades of grey. The monitor is sometimes also known as screen, display or a CRT (Cathode Ray Tube). Images produced on a computer depend on the screen used. An image is made up of tiny dots called pixels. Latest development in monitor is LCD (Liquid Crystal Disply) or TFT

TET Monitor

CRT Monitor

Printers

Printer is a device that produces written images on the paper. After a document had been created on a computer it can be sent to a printer for a hard copy, known as a print out. The speed of the printer is rated either by pages per minute (PPM) or by characters per second (CPS). The more the dots per inch, the more details is the output. Printers come with many different shapes, sizes and printing technologies. The commonly used printers are (a) Dot matrix printers, (b) Ink-jet printers, and (c) Laser printers.

Dot Matrix Printer

A dot matrix printer uses tiny dots to create text and graphics on paper. The print head contains 7 to 27 pins. It constructs a character by repeatedly striking these pins against an inked ribbon and the paper.

Ink jet Printer

An ink-jet printer has a mechanism that sqirts tiny electrically charged droplets of ink, out from a nozzle and onto the paper. No pins or hammer strikes the paper. They are fast and quiet and can produce color prints also.

Laser Printer

A laser printer uses highly focused beams of light to transfer images on the paper.

Laser Printer

Mouse

Mouse is a small object, designed to fit neatly under palm. It has two main functions, one can use it to move the cursor on the screen in all directions (Left, Right, Up, Down or Diagonal), and also select the items displayed on the screen. The most useful feature of the mouse is its buttons which help us in selecting items. Generally a mouse has two buttons, while some types contain three buttons.

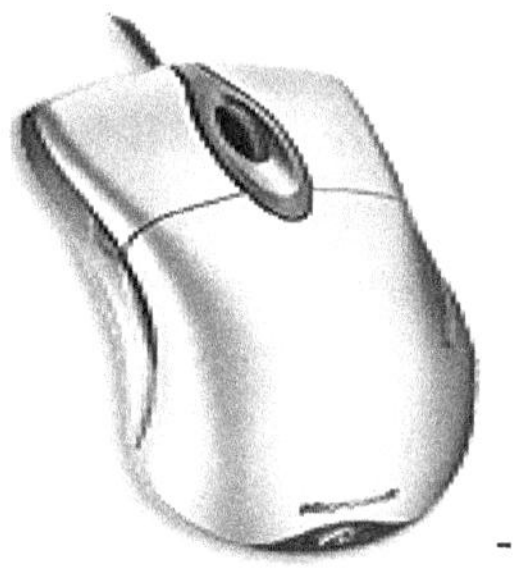

Optical mouse

Key Board

Esc: Escape	:	It escapes the message box
F1 to F12	:	These are called Function Keys
`!@#$%^	:	These are called Special keys
Tab	:	When pressed it displays all Capital Letters.
Shift	:	To get the starting letter Capital press Shift+ Key.
Ctrl	:	Control Key. It does not work independently we have to use another key; example: Ctrl + C for Copy.
Window Key	:	To display the Menu it is useful

Key Board

Alt key	:	It does not work independently we have to use another Key; example: Alt + F4 for close a window

Properties Key	:	It displays the additional properties which it contains. But first You have to select it. And press this key.
Space Bar	:	It allows single space between the words.
Enter	:	To change for second line or give the command we use this key.
Backspace	:	To go backward direction we use this key. But it also deletes words which ever is behind the cursor.
Alphabetic keys	:	It displays each alphabets.
Insert	:	To insert any specific word we use this key. After the use again we have to press it once to de-activate its action.
Delete	:	To delete the word which are ahead of the blinking cursor and also to delete the files /folders we use this key.
Home	:	To go for the starting point.
End	:	To go for the ending point.
Page up	:	It moves the page in the upward direction.
Page down	:	It moves the page in the downward direction.

Cursor Control keys

These keys enable you to move the cursor to the left, Right, Up and Down Arrow Keys respectively. These keys enable you to move the cursor to the left, right, up or down on the screen- one character at a time.

Num Lock

It should be always "ON" to display Numeric value. When it is "OFF", then Home, Up arrow, Page Up, Left arrow, Right arrow, End Down arrow, Page down, Insert and Delete options will work.

CD /DVD ROM DRIVE

CD / DVD

It is Compact Read only drive, reads data on surface of CD. CD Writer will record the digital data onto the Compact Disk (CD). CD stores the data in optical form. It can store maximum data of around 700 MB.

CD SPEED   : *Reading speed* varies from 1x (150kb/s) to 52x (7800 kb/s).

Writing speed available is 4x, 8x, 16x, 24x, 32x, 48x, 52x.

Development after CD is DVD (Digital Versatile Disc). In general, capacity of a DVD is 4.7 GB. However, capacities ranging from 8.75 to 20, 40 GB etc., are also available.

DVD SPEED   : *Reading Speed* varies from 1x (1385kb/s) to 16x (22160kb/s). *Writing Speed* available is 4x, 8x, 16x only.

Depending upon the data present on the DVD surface, DVD Rom reads from 1x to 16x. 20 GB, 40 GB, DVDs can be read by blue ray DVD Rom.

PC ASSEMBLING

- ➤ Connecting various part of a computer is called the PC assembling
- ➤ P-IV 1.6 GHz 478 pin CPU
- ➤ Mother Board
- ➤ SD RAM
- ➤ FDD
- ➤ CD ROM Drive
- ➤ Hard Disk Drive

STEP BY STEP PROCEDURE OF ASSEMBLING VARIOUS COMPONENTS

1. Preparing Chassis

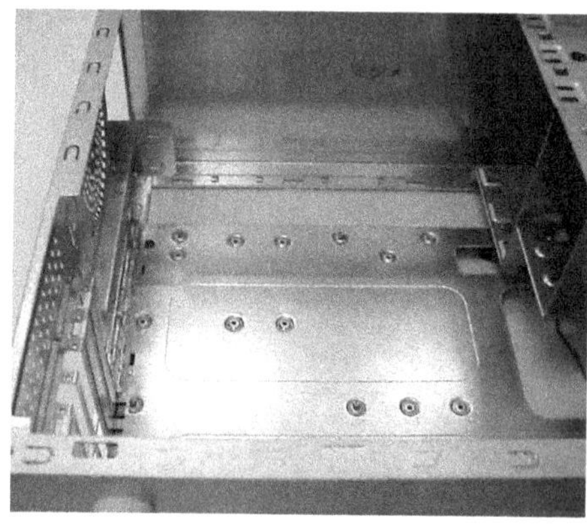

Empty Cabinet

2. Installing Mother Board

a. Inside CPU

b. Rear view of CPU showing various ports

3. Installing CPU

a. Mother board Processor Location

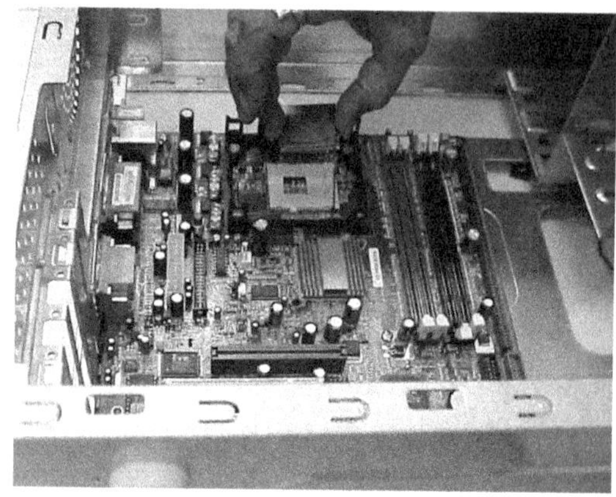

b. Placing Processor Base

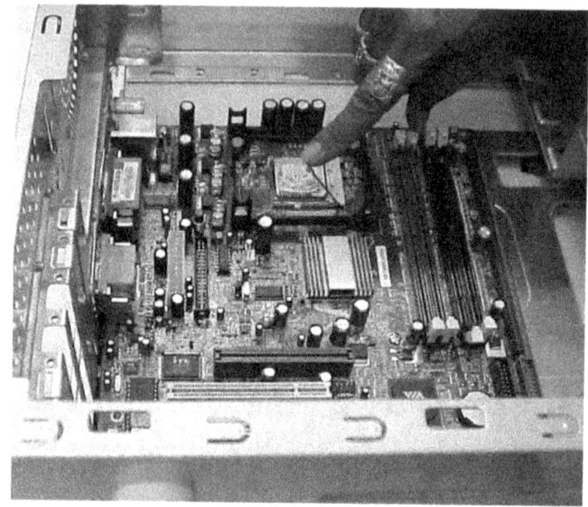

c. Locking Processor pins

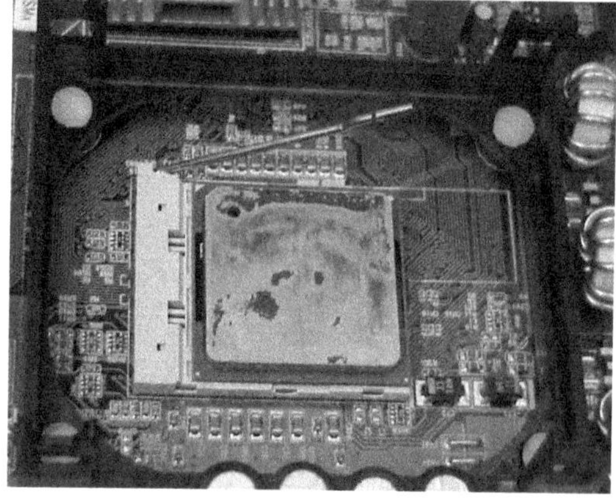

d. Processor before Locking

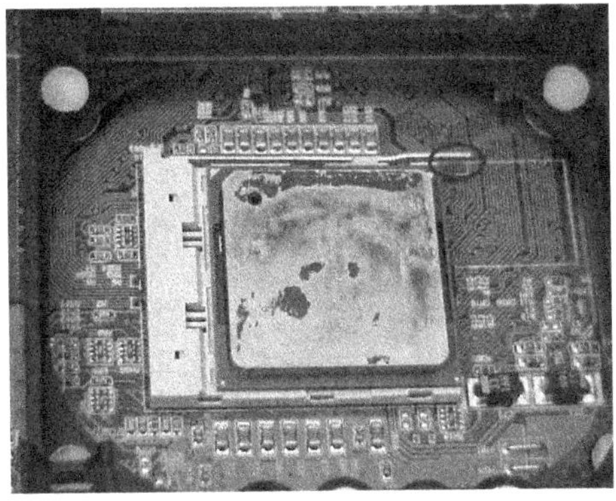

e. Processor after Locking

4. Installing CPU Heat Sink and Fan

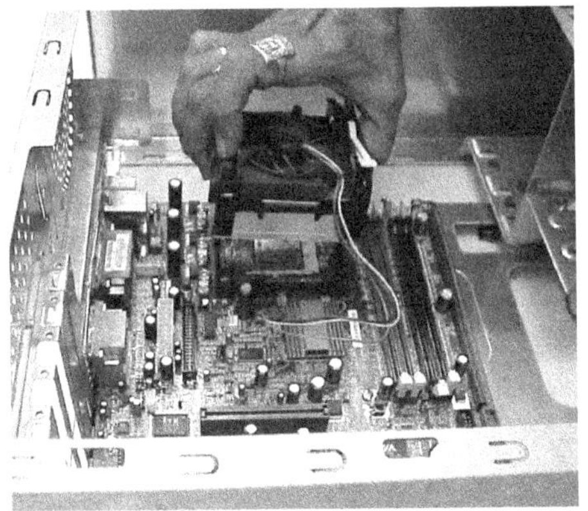

a. placing CPU Heat Sink and Fan

b. View after placing heat Sink and Fan

c. Power Supply to the Processor Fan

5. Installing RAM

a. Inserting RAM into the respective DIMM Slots

(i.e., SD RAM or DD RAM Slots)

b. Locking RAM

6. Installing SMPS

a. Inserting SMPS

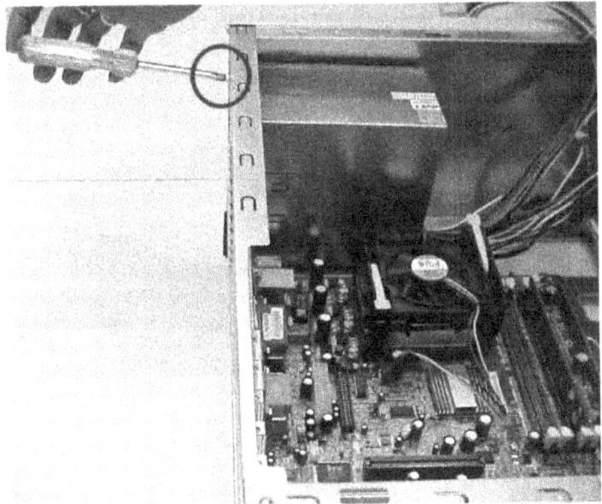

b. Fixing SMPS

7. Installing ATX Power Connector

c. Power connection to Mother Board

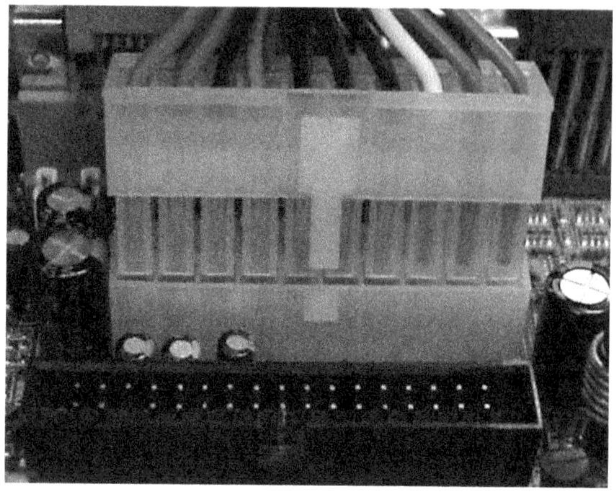

d. Closer View of Mother Board power supply

8. Installing The Hard Disk Drive

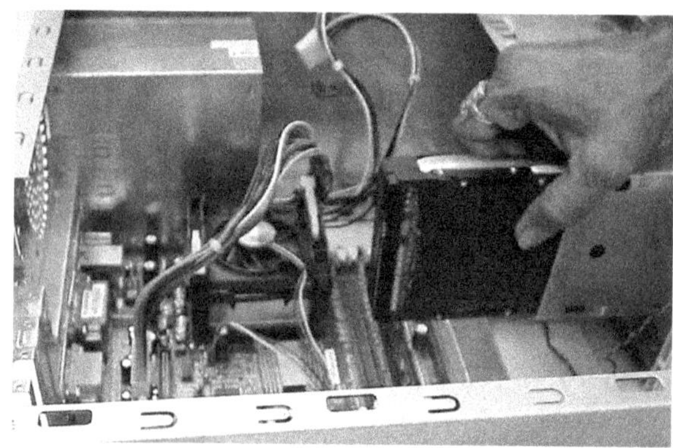

a. Inserting HDD

b. Fixing HDD

c. Connecting Bus with IDE-1 in the Mother Board

d. Closer view of IDE-1 slot

e. Bus connection to HDD

9. Installing FDD

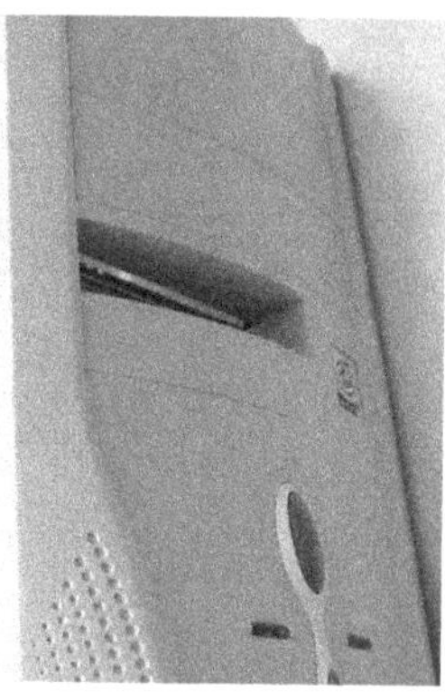

a. Location of FDD in the Cabinet

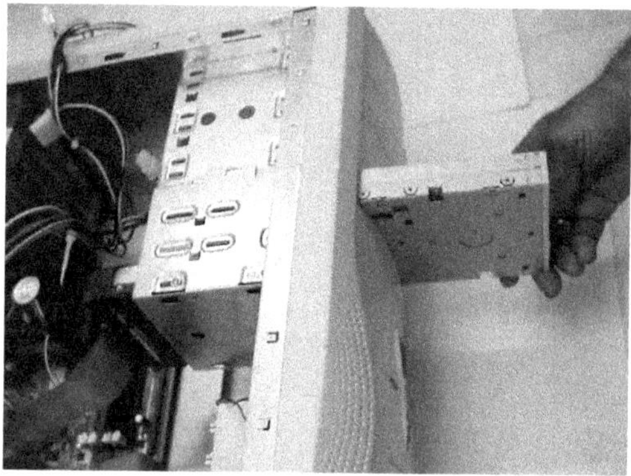

b. Inserting FDD in the Cabinet

c. View After Inserting FDD

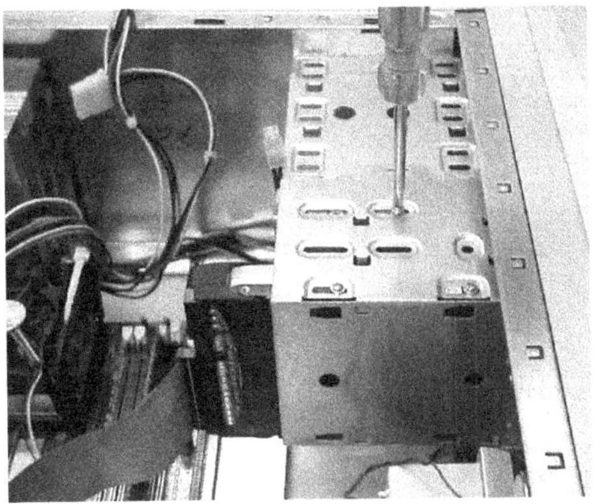

d. Fixing the FDD

e. Connecting FDD Bus into the Mother Board IDE Slot

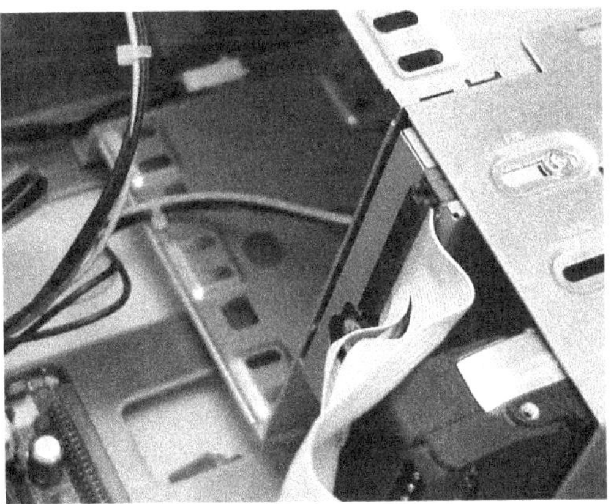

f. Connecting Bus to the FDD

10. Power Connection to HDD and FDD

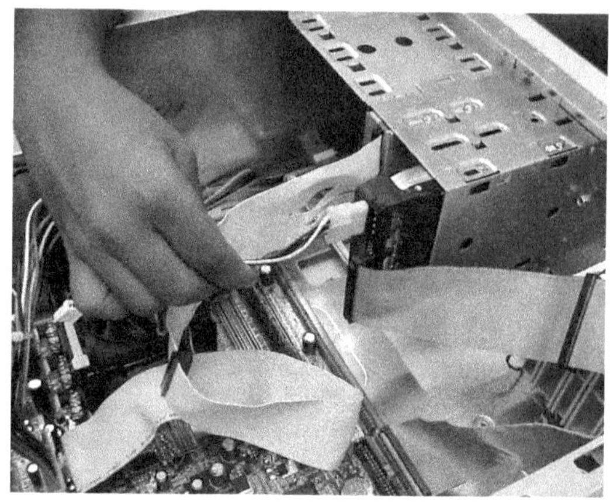

a. Connecting Power Supply to HDD

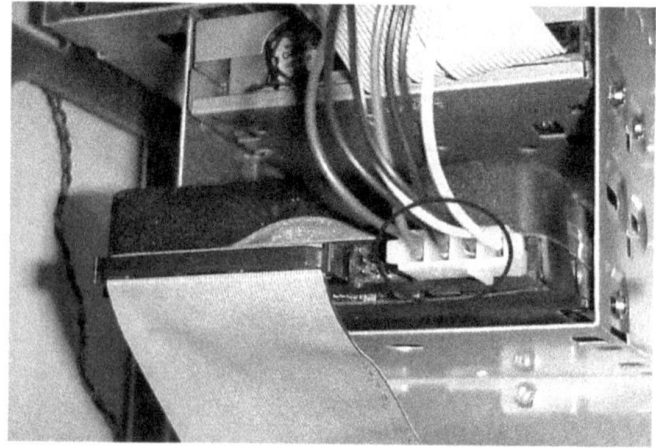

b. Closer view of HDD Power Supply

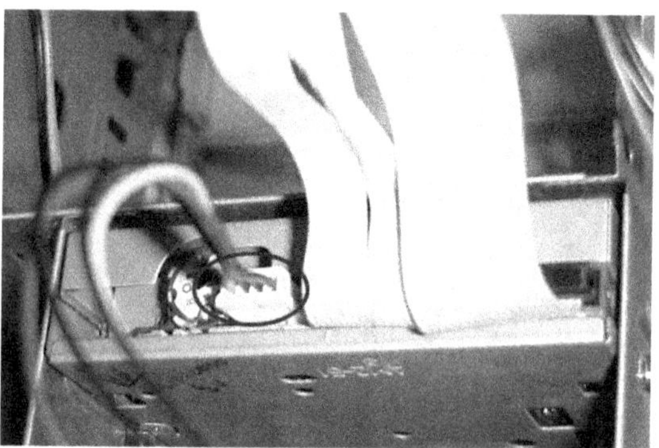

c. View after connecting Power Supply to FDD

11. Installing CD/DVD ROM Drive

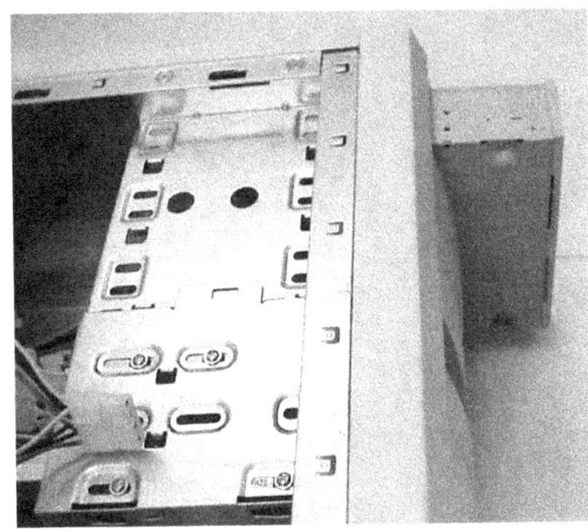

a.   Inserting CD/DVD ROM

b. After Inserting CD/DVD ROM

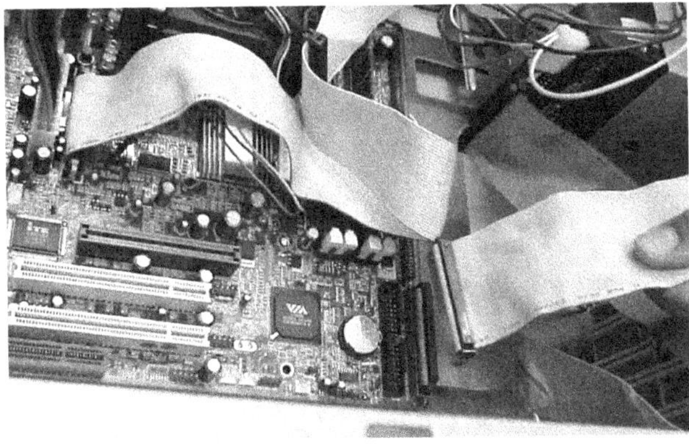

c. Connecting the Bus to the IDE-II Slot in Mother Board

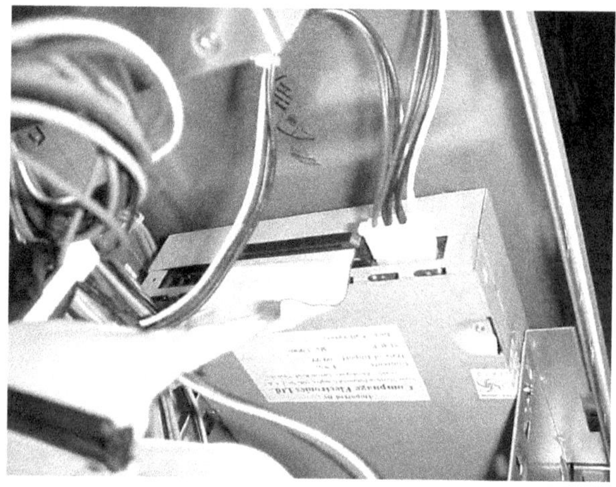

d. Power Supply and Bus to the CD/DVD ROM

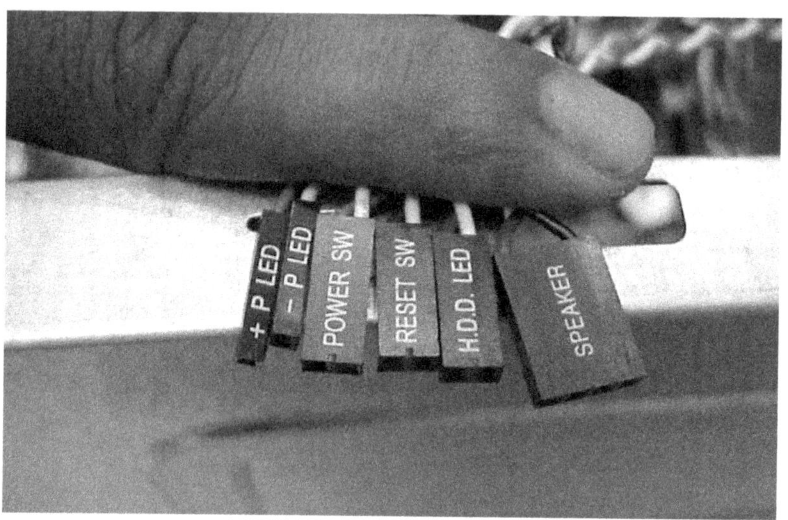

LED and Speaker Connections to be Inserted into the Mother Board

Guide lines to be followed while purchasing a system

Configuration varies depending upon the user of application. For example applications like SAP, MAYA etc., require highest configuration. One should concentrate on processor, RAM. Such type of configuration is given below.

1. Processor   :   Intel core2duo (2.8 GHz)
                         Bus Speed 800 MHx – Data transfer rate
                         2 MB L2 CACHE (CAHE memory)
2. RAM   :   1GB DDR2(800MHz).

3. Mother Board   :   975 GNT Intel.

4. Hard disk   :   160 GB serial ATA hard drive
                               Data transfer rate 150 Mb/sec@7200 rpm.

5. Optical disk   :   DVD 18 × R/writer

6. Multimedia key board & optical mouse.

7. ATX Cabinet & SMPS 400 watts

8. LCD (or) TFT Monitor (17"m 19" etc).

9. UPS

The above configuration is given for higher order applications. However, it varies depending upon the purpose of application. For Example multimedia applications AMD authelon processor is suggested.

For games applications we should concentrate on Graphic Card. It is a PCI card, which is inserted in PCI slot of motherboard. It has extra memory. If it is built in the motherboard (and Graphics card is not inserted) it will share RAM memory. For example, System Ram of 1GB, Built in Graphics card will share 110MB approximately.

Note : Hard disk partition must be in NTFS format. NTFS format will consume some amount of hard disk memory (for Ex : 160GB, 10GB is utilized). However, it is not with FAT format. NTFs format is preferable because of its high security comparatively. Therefore, Server Systems must be in NTFS format.

Chapter - 11

POWER TOOLS

INTRODUCTION

Portable power tools are widely used in various fields of engineering. There are many manufacturers supplying the same with wide range of products. One should be careful in selecting the tools. Double insulation tools are recommended for safety. Selection of range or capacity is also important depending on the application. The following are some of the fields of use.

(a) Simple drilling operation, (b) Hammer drilling in hard building materials and concrete, (c) Carpentry work involving (i) planning, (ii) cutting (iii) Sander, etc. (d) Marble cutter, (e) Various types of grinders, (f) Blower, (g) Screw driver etc.

1.   **Drills (wood and metal):** These are used for drilling holes in wood/metal.

Capacity:
       Steel : 10 mm
       Wood : 25 mm
Power Input: 420 W
No-load Speed: 2,800 rpm
Weight: 1.4 kg

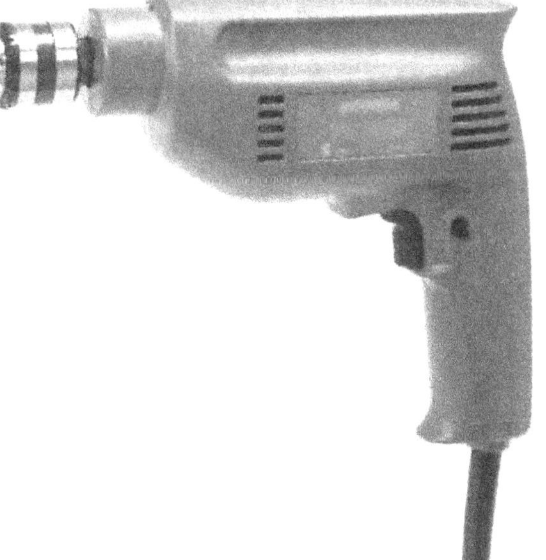

2. **Impact drills (concrete and brick):** These are used for making holes in concrete and bricks. Impact drills with multipurpose application are used for general drilling as well as impact drilling.

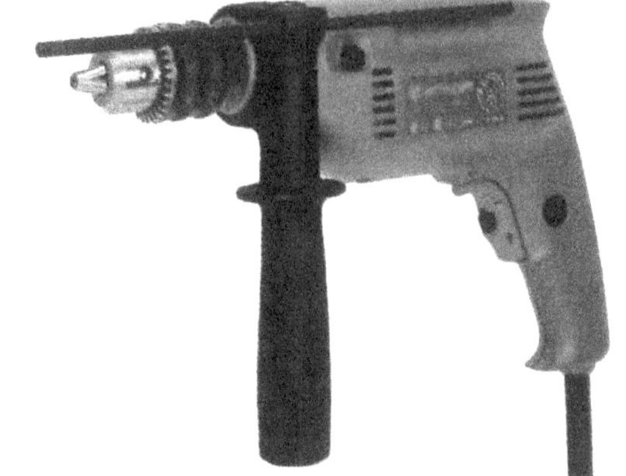

Capacity:
    Concrete : 15 mm
    Steel : 10 mm
Power Input: 520 W
No-load Speed: 0 ~ 2,900/min
Impact Rate: 0~46,400/min
Weight: 1.6 kg

3. **Hammer drills (concrete and brick):** These are heavy duty drills for making holes continuously with bigger size drills up to 38 mm. It has keyless drill check.

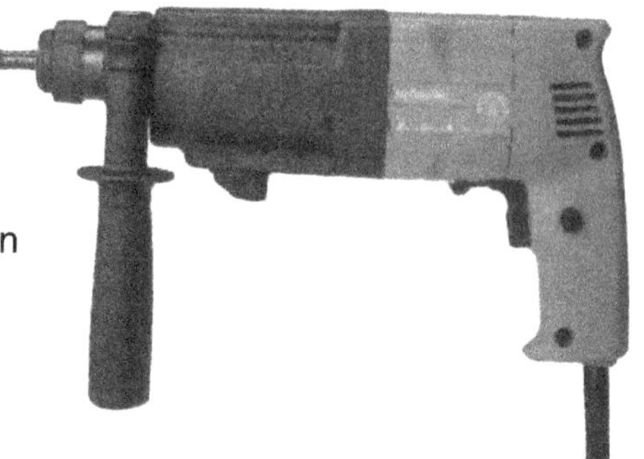

Capacity:
    Concrete : 24 mm
Power Input: 500 W
No-load Speed: 0~1000 blows per min
Weight: 2.3 kg

4. **Masonry cutter:** These are used for cutting to size floor tiles, granite stone slabs or tiles etc.

Capacity: 110 mm (Wheel dia)
Max. Cut. Depth: 34 mm
Power Input: 1,050 W
No-load Speed: 11,000 rpm
Weight: 2.6 kg

5. **Circular saw (wood):** This is used to cut wood.

   Capacity: 185 mm (Blade dia)
   Max. Cut. Depth:
       65 mm (90˚)
       43 mm (45˚)
   Power Input: 1,050 W
   No-load Speed: 4,700 rpm
   Weight: 3.9 kg

6. **Compound mitre saw (wood):** This is used for precision cutting and any angle cutting.

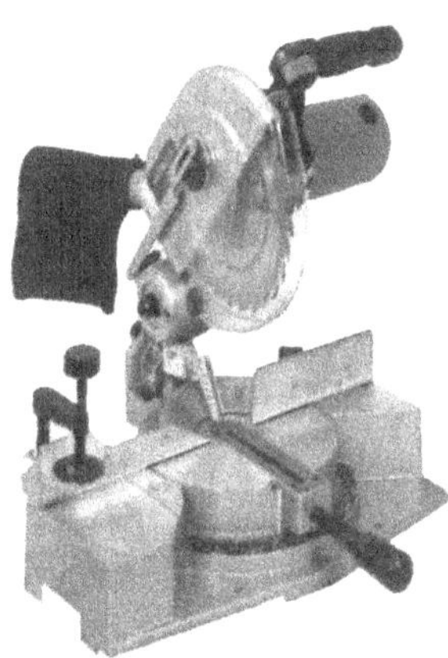

   Capacity: 255 mm (Blade dia)
   Max Cut. Depth: 0° Cross Cut 67 × 145 mm
       Milter 45° 70 × 89 mm
       Bevel 45° 44 × 130 mm
   Compound: $45^\circ \times 45^\circ$ 44 × 89 mm
   Power Input: 1,640 W
   No-load Speed: 5,000 rpm
   Weight: 14 kg

7. **Jig saw (wood and metal):** This power tool is used to cut along a line. It can also be used for bevel cutting 0-45˚ right/left.

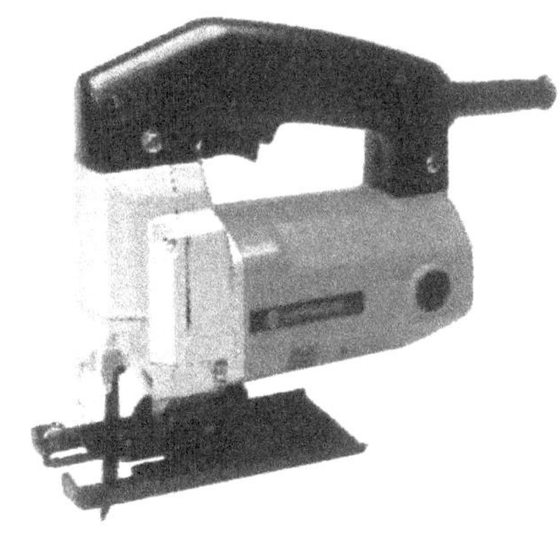

   Max. Cut. Capacity:
       Wood: 600 mm; Steel: 6 mm
   Bevel cutting: 0 ~ 45˚ (R/L)
   Power Input: 400 W
   No-load Speed: 2,700/3,200 rpm
   Weight: 2.1 kg

8. **High speed cut-off machine (metal):** This is a versatile tool to cut metal with speed and accuracy. Very much in use to cut torsion rods in building construction at site.

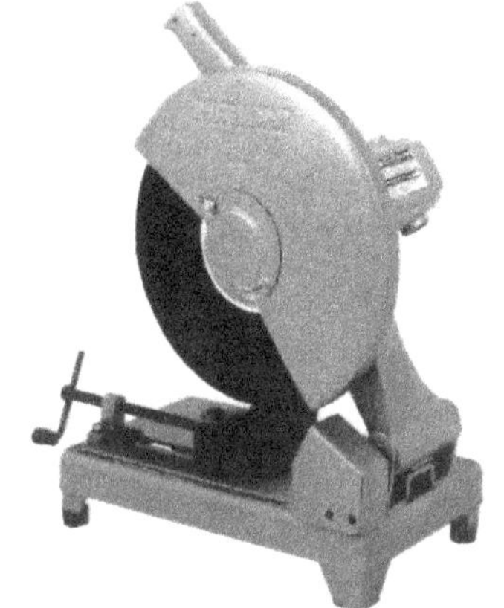

Capacity: 355 mm (Wheel dia)
Max. Cut. Capacity:
    Pipe: 114.3 mm
    Bar : 65.0 mm
    Shape steel: 130 × 130 mm
Power Input: 2,000 W
No-load Speed: 3,700 rpm
Weight: 17.5 kg

9. **Chain saw (wood):** This is used for continuous sawing of wood. Mostly used for felling trees.

Capacity: 350 mm (Bar size)
Power Input: 1,140 W
No-load Speed: 450 m/min
Weight: 4.2 kg

10. **Orbital sander (metal and wood):** This versatile tool is used for smooth finishing of wood and metal surfaces which is given orbital action for fast and smooth finish on plain surfaces.

Capacity: 110 × 100 mm (Pad size)
Power Input: 180 W
No-load Speed: 24,000/min.(Orbit)
                  12,000/min.(Spindle)
Weight: 1.1 kg

11. **Disc sander (wood):** Used to give smooth finish to wood surface with sand discs mounted and rotated in circular motion unlike orbital sander. It can reach irregular shapes and spaces.

Capacity: 150 mm (Sanding Disc.)
Power Input: 530 W
No-load Speed: 4,000/min
Weight: 3.2 kg

12. **Sander polisher:** This is used for polishing wood surface.

SAT-180
2 Speed for Various Application
Capacity: 180 mm
Power Input: 750 W
No-load Speed: 1,900/3,400/min.
Weight: 2.9 kg

- Two application: Sanding & Polishing
- Option: With or Without Wool Bonnet

13. Disk grinder (metal): Used for grinding metal to give required shape and finish especially after welding.

Capacity: 115 mm (Wheel dia.)
Power Input: 750 W
No-load Speed: 10,000/min
Weight: 2.1 kg

14. Hand grinder (metal): This is a versatile power operated portable grinder which can be used in number of ways for both internal and external grinding of various conturs by suitably changing the mounted chuck.

For Precise Grinding
Capacity: 6 mm (Collect Chuck)   25 mm (Max. Wheel dia.)
Power Input: 520 W
No-load Speed: 25,000/min
Weight: 1.8 kg

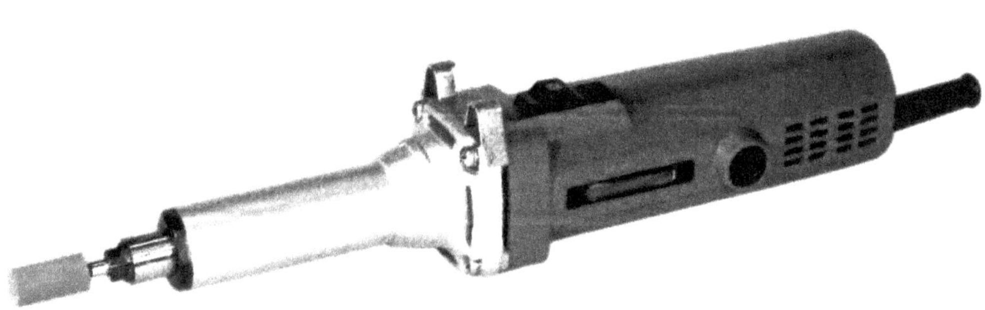

• Wide selection of Mounted Wheel

15. **Planer (wood):** This is used for power planning of wooden surfaces in the place of hand jack plane.

Capacity:
    Cutting Width: 82 mm
    Cutting Depth: 1 mm
Power Input: 570 W
No-load Speed: 15,000/min
Weight: 2.5 kg

16. **Screw driver:** This particular tool is used to drive screws in wood or sheel metal. It can also be used for removing the screws by reversing the rotation.

Capacity:
    Wood Screw: 5.8 mm
    Drywall Screw: 5 mm
    Self Drilling Screw: 6 mm
Power Input: 520 W
No-load Speed: 0 ~ 2,600/min
Weight: 1.7 kg

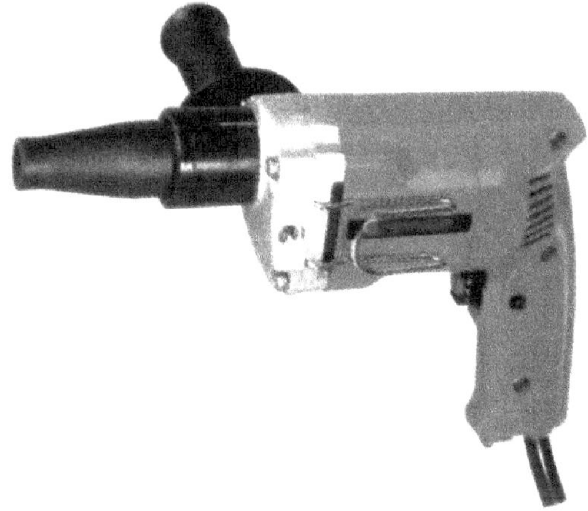

17. **Blower:** Blowers are used for cleaning and maintenance with dust bags to collect the dust.

Capacity:

Air pressure: 400 mm

Air volume : 2.3 m^3/min

Power Input: 355 W

No-load Speed: 13,000/min

Weight: 1.8 kg